INTERCROPPING SYSTEM

NIPA® GENX ELECTRONIC RESOURCES & SOLUTIONS P. LTD.
New Delhi-110 034

Intercropping System Theory and Practices

Sagar Maitra, Ph.D.
Professor and Head
Department of Agronomy and Agroforestry
Coordinator, Research Center for Smart Agriculture
M S Swaminathan School of Agriculture
Centurion University of Technology and Management, Odisha

NIPA® GENX ELECTRONIC RESOURCES & SOLUTIONS P. LTD.
New Delhi-110 034

NIPA® GENX ELECTRONIC RESOURCES & SOLUTIONS P. LTD.

101,103, Vikas Surya Plaza, CU Block
L.S.C.Market, Pitam Pura, New Delhi-110 034
Ph : +91 11 27341616, 27341717, 27341718
E-mail: newindiapublishingagency@gmail.com
www: www.nipabooks.com

For customer assistance, please contact
Phone: + 91-11-27 34 17 17
Fax: + 91-11-27 34 16 16
E-Mail: feedbacks@nipabooks.com

ISBN: 978-93-94490-42-0

Composed and Designed by NIPA®.

To
My father
An eminent scholar
Late Jyotishwar Maitra
Who shaped my life

Bidhan Chandra Krishi Viswavidyalaya
P.O. Krishi Viswavidyalaya, Mohanpur 741252
District: Nadia, West Bengal, India
Website: www.bcky.edu.in

Phone: 03473299918
Email: vc@bcky.edu.in
bckvvca gmail.com
Cell: +91-9830071278

Prof. B S Mahapatra
Vice-Chancellor

No. VC/BCKV/87
Date : 17.03.2022

Foreword

Concurrent agriculture is facing tremendous problems with dwindling natural resources and diversity, depletion of soil, water and air quality along with decreased factor productivity and net revenue. All these issues caused threats to the food, nutrition and livelihood security of the farmers as well as the human population of the world. The situation of small and marginal farmers in the developing countries is comparatively gloomy because of lesser investment capability of the farmers and adoption of subsistence farming. Moreover, global warming and climate change created another dimension into the problems.

In the present consequences, the small and marginal farmers need to adopt a suitable cropping system which is ecologically sound, remunerative, efficient in utilizing natural resources, effective for achieving food and nutritional security, and agricultural sustainability. In this regard, a suitable intercropping system fit to a particular agroclimatic zone and cropping season can be considered as a viable option.

Growing two or more crops in the same field and their coexistence for a sizable period of their lifecycle is termed as intercropping. However, the age-old intercropping system has enough potential in the present context even to combat the ill effects of global warming as mitigation and adaptations. In general, intercropping plays a multifaceted beneficial role such as diversity in food basket, higher net return from unit area and production stability, improvement of soil quality, proper utilization of human resources leading to sustainability in agriculture.

I congratulate the author for bringing out the book "Intercropping System" which consists of valuable inputs in cropping systems with the theory and practical example. I believe that the book will be useful to the undergraduate and postgraduate students, researchers and agronomists. I wish the success of the book and appreciate the efforts of the author.

(B. S. Mahapatra)

My Words

Modern agriculture introduced high-yielding, fertilizer-responsive and photo-insensitive varieties and hybrids of crops. These were cultivated with enough of different inputs, namely, synthetic chemical fertilizers, plant protection chemicals and assured irrigation which led to a breakthrough in agricultural production. However, the small and marginal farmers of resource-poor regions were not able to acquire the benefits of these technologies because of less investment and ecological constraints. Modern agriculture could not accomplish the goal of sustainability; rather, food grain production was enhanced at the cost of agro-ecosystem. On the other hand, smallholders of the developing countries still practice a diversified cropping system.

Over the years, it has been realized that the adoption of a suitable cropping system is a prerequisite for agricultural sustainability. Multiple cropping can be adopted mostly under assured irrigations, but in rainfed and dryland conditions there are limited options for crop diversification. It may be noted that small farmers of drylands are familiar with adopting polyculture in the form of intercropping for their subsistence and livelihood. Intercropping was considered as a poor farmer's activity because of some limitations such as involvement of more labourers, less scope of mechanization, allelopathic effects among the crops grown together and so on. During the present time, ill effects of modern agriculture are pronounced as yield plateauing, degradation of resources, environmental pollution and adverse impacts of climate change. Under these adverse conditions, researchers, agriculturists and policymakers are in search of an ecologically sound option and thus, intercropping has gained renewed attention. Interestingly, intercropping is a felt need of the hour and it has the qualities to overcome all the issues caused by the adoption of industrialized modern agriculture; and thus, is recognized as an ideal system for agricultural sustainability.

During recent times, the research dimension on intercropping has changed and innovative findings are distinguished favouring multifaceted benefits. However, owing to having scarcity of comprehensive textbooks on intercropping inspired me to pen down this book addressing the advantages and management of the intercropping system. Students, researchers and policymakers can get necessary inputs for the adoption of a suitable cropping system through this book. The book comprises nine chapters covering all the technological details and versatile impacts of intercropping on sustaining agricultural productivity.

Completion of this book was a mammoth task as I am not an inborn researcher, rather a developmental functionary. The research in my academic career and outstanding features of intercropping forced me to write about the same. Writing of the book would not have been possible without the help and encouragement from my well-wishers. First of all, I must pronounce the name of Prof. D. N. Rao, Vice President of Centurion University of Technology and Management who supported me in my professional career. I am grateful to Prof. Supriya Pattanayak, Vice-Chancellor of Centurion University and Technology and Management for her constant inspiration. I am indebted to Prof. Anita Patra, Registrar, Centurion University of Technology and Management for her support. I must recognize the enthusiasm provided by Dr. Dipankar Bhattachayay, Pro-Vice-Chancellor (Research and Learning) of Centurion University of Technology and Management. I acknowledge the care received from Dr. S. P. Nanda, Dean of Administration and Dr. M. Devender Reddy, Dean of M.S. Swaminathan School of Agriculture. Further, Dr. Tanmoy Shankar, Dr. Subhasisha Praharaj, Dr. D.T. Santosh, Ms. Jnana Bharati Palai, Mr. Masina Sairam and Mr. D. Sarath helped me a lot during writing the book. I also deeply acknowledge my gratitude to Dr. Parthendu Poddar, Professor of Agronomy, Uttar Banga Krishi Vishwavidyalaya and Dr. Biswajit Pramanick, Assistant Professor, Dr. Rajendra Prasad Central Agricultural University who have gone through the manuscript and provided their valuable suggestions for improvement of the book. I express my heartfelt gratitude to Mr. Sekhar Maitra and Mr. Sandipan Pine for their continuous encouragement. I must mention the name of Mr. Balai Lal Jana, a brilliant scholar and author of several books in agriculture, who provided enthusiasm during the major period of my professional career. I am very much grateful to Dr. D.C. Ghosh and Dr. G. Sounda, retired Professors of Visva-Bharati and Bidhan Chandra Krishi Vishwavidyalaya, respectively who taught me the basic knowledge of intercropping systems.

I am particularly thankful to Mrs. Shrabanti Maitra for her devoted efforts in the development of necessary figures and continuous support in guiding me to give a proper shape to the book. I must acknowledge my thanks to Mr. Samantar, Mr. Shubhankar and Ms. Swanka whose affections were with me during the process.

I convey my gratitude and regards to Prof. B. S. Mahapatra, Hon'ble Vice-Chancellor of Bidhan Chandra Krishi Vishwavidyalaya who encouraged me in writing the book and made me ever-indebted by penning down the valuable foreword of the book.

I am also thankful to the management of the NIPA for their sincere initiatives in publishing the book.

I took around three and half decades to write a book on the intercropping system since I carried out my first research work on the topic in 1994. I tried to explain the intercropping system based on the knowledge I gained. I shall be happy enough if the book helps the stakeholders. All criticisms will be highly solicited.

May 16, 2022

Sagar Maitra
Paralakhemundi, Odisha

Contents

1

Overview of Intercropping

Predominantly, an agrarian society is mostly dependent on the farm sector. Agriculture plays an important role in Indian economy notwithstanding its shrinkage in share in the gross domestic product (GDP) of the country being 55.4% in 1950-'51 and 17.4% in 2018-'19 (Tripathy, 2019). Still the agriculture and allied sector is providing employment to 54 percent of the Indian population. The prime target of the current agriculture front in India is to enhance farm outputs for growing needs of the present as well as future from the declining and degrading arable land. The land shrinkage for the farming sector is because of its non-agricultural uses as driven by urbanization and industrialization. Under this circumstance, implementation of intensive cropping systems can be a choice to enhance the farm output and optimum income from the farming sector (Maitra *et al.*, 2019).

Agriculture is a ritual in India and so many countries and it has a glorious legacy. The agriculture of earlier days was with some 'so-called' outdated practices of farming in which mixed stands of various crops were prioritized which were nothing but the mixed or intercropping. Further, the concept of the farming system adopted since ancient times in the different nooks and corners of the world can also be considered as poly-culture or mixed culture (Plucknett and Smith, 1986). The early human civilizations evidenced the practice of intercropping (Papanastasis *et al.*, 2004) in different forms. In the tropical regions of the world, intercropping is practiced with mixed stands of food crops. However, in temperate countries intercropped forage and cover crops cultivation is a common practice (Anil *et al.*, 1998; Maitra *et al.*, 2021).

Traditional agricultural practices demonstrated worldwide with a mixed culture of crops can be considered as mixed or intercropping. Early civilizations experienced cultivation of crop mixtures in various forms (Plucknett and Smith, 1986; Fuller, 2011). During the era of the Indus valley civilization, different types of intercropping or mixed cropping were practiced. The archaeologists recorded that at the *Kalibangan* (Rajasthan) site, there was evidence of adoption of mixed cropping (Lal, 1970). Intercropping was a common practice in China also during the ancient period and historical evidence narrated that it was initiated thousand years back (Knörzer *et al.*, 2009). During the ancient times (300 BC), the practice of intercropping of pulses and cereals (barley and wheat) in Greece was recorded (Wright, 2010). In the writings of Theophrastus, example of intercropping was narrated in the early Greek civilization (Wright, 2010). Globally, traditional mixed culture of crops contributes about 15-20% of foods and that too is concentrated in the developing countries (Altieri, 1999). Intercropping is presently practiced in Chinese agriculture (137 million ha) covering a major part in the form of agro-forestry (Li *et al.*, 2007). But in China, relay intercropping and strip intercropping are common. However, about 17% of arable land in India is engaged in intercropping (Knörzer *et al.*, 2009). In African countries also, a considerable area is noted under intercropping (Knörzer *et al.*, 2009). Maize and legume intercropping is very popular cropping system in Latin America (Francis, 1986) and in Columbia about of 90% beans cultivated as intercrop (Maitra *et al.*, 2021). In the African countries, the majority of cowpea is produced as intercrop and in Malwai 94% of the country's arable land is under intercropping system (Vandermeer, 1989). In Ethiopia, 91% cropping system is comprised of intercropping (Fininsa and Yuen, 2001) and maize + bean intercropping is very common in East and Southern Africa (Mucheru-Muna *et al.*, 2010).

Modern technology based industrialized agriculture is no-doubt easy to the resourceful farmers where sufficient investment in agriculture is observed and timely supply of necessary input as well as provision of farm machinery are ensured. Such farmers can get easily connected to the market. But there are a large number of farmers dwelling in the developing countries who produce crops with little investment under sub-optimal conditions. They cannot access the benefits of modern monoculture and for them intercropping can be considered as an automatic choice for the food and nutritional security with livelihood improvement. By enriching the diversity of the crop basket, intercropping ensures yield stability and environmental safety. There is evidence that intercropping maintains the population dynamics of insect-pest, weeds and pathogens and therefore, in organic agriculture, intercropping adds a great value as an important cropping system (Bulson *et al.*, 1997; Jensen *et al.*, 2005; Maitra and Gitari,2020).

During the present time, system approach in agriculture has got priority among the researchers and policy makers. In a system, the components present there remain related and interaction among them is observed. Therefore, managing a system is rather more important and effective than management of individual crops grown in the system. Further, the system approach focuses on the fullest utilization of resources available targeting an increase in cropping intensity with agricultural sustainability. An intensive cropping system is comprised of cultivation of more number of crops and varieties with the desired resistance or tolerance to abiotic and biotic stresses. Moreover, maintenance of soil fertility and enhancement of production from the unit area are also prime targets of an intensive cropping system. A suitable cropping system for an agro-climatic zone can ensure target yield and in this regard, available resources and their management options play vital roles for evolving the cropping system. Finally, the efficiency of a cropping system greatly depends on net area under cultivation, crop yields and duration of the system (Willey and Reddy, 1981; Willey *et al.*, 1983).

Undoubtedly, modern agriculture boosted crop productivity, but ecological matters were not considered (Tilman *et al.*, 2002). Moreover, in modern agricultural practices, crop diversity was not considered and focus was given on yield enhancement of few crops and their varieties/ hybrids which resulted in genetic erosion. On the other hand, intercropping or poly-culture nurtures on-farm biodiversity with cultivation of various crops, efficient use of resources and engagement of human workforce (Anil *et al.*, 1998). In intercropping, all the crop species grown together may or may not be sown and harvested simultaneously, but they coexist in the field for a considerable period. In general, physiologically and morphologically different types of crops are chosen in an intercropping system.

A crop production system with low input involvement but simultaneously energy-efficient is always desirable for small farmers (Altieri, 1999). If the system ensures agricultural sustainability, it can easily be adopted (Jackson *et al.*, 2007; Scherr and McNeely, 2008). An intercropping system has all the advantages. However, in industrialized modern agriculture, high energy synthetic inputs are used causing harm to the agro-ecosystem as well as creating issues for achieving agricultural sustainability (Tilman *et al.*, 2002; Lichtfouse *et al.*, 2009). In India, agricultural production was increased drastically after the Green Revolution and the country witnessed its ill effects within few years; whereas, intercropping is a system consisting of all positive aspects for achieving agricultural sustainability such as crop diversity, ecological balances, low-cost and energy efficiency, proper resource utilization and soil fertility restoration.

1.1. Concept and Goal of Intercropping

Growing two or more crops simultaneously in the same land is termed as intercropping or mixed cropping or poly-culture (Andrews and Kassam, 1976). As mentioned earlier that in intercropping system, the component crops remain on the field for a considerable period of the crop cycle of component crops species. Generally, intercropping is consisting of a main crop and one or more companion crops, where focus is concentrated on the productivity of the main crop. But where both the crops are equally important, they are considered as component crops. By adopting an intercropping system, the value of the cropping system can be added with a target of productivity enhancement, better utilization of resources and achievement of economic benefits (Maitra *et al.*, 1999; Maitra *et al.*, 2000; Manasa *et al.*, 2018). In spite of versatile benefits of intercropping, the focus was given on modern agriculture for achieving the goal of more production and the ill effects were also realized within a decade or two after the Green Revolution in India. The increasing requirement of agricultural sustainability and reduction of the anthropogenic activities in environmental degradation reinforce to revisit the multifaceted benefits of intercropping.

Under the mixture of diversified crop species spatially and temporally, intercropping is recognized as an agro-ecological practice (Rodriguez *et al.*, 2020). When the complementary effect among the crops is pronounced, the maximum advantages of intercropping are harnessed (Maitra *et al.*, 2020). The above and below-ground interaction among the crops plays a vital role (Lithourgidis *et al.*, 2011a; Bedoussac *et al.*, 2015) in expression of the advantages and therefore, crops are chosen accordingly (Maitra *et al.* 2019). The below-ground interaction is more pronounced when legumes are chosen in mixed stands with non-legumes in terms of nitrogen transfer (Maitra *et al.*, 2020), phosphorus availability (Li *et al.*, 2009) and enrichment of soil microbial diversity (Maitra and Ray, 2019). All of these are ultimately reflected into yield stability by the compensation effect (Rao and Willey, 1980). Besides, in intercropping, various resources related to crop production such as space, water, light and nutrients are efficiently used as compared to the sole cropping of the respective crops resulting in higher combined yield (Li *et al.*, 2006; Bedoussac *et al.*, 2015).

In the present consequence of climate change and a huge load of carbon footprints in agriculture, the intercropping system may be reevaluated considering its numerous advantages. The United Nations announced that the Sustainable Development Goals (SDG) is to be achieved by 2030. Intercropping offers multifaceted advantages such as enhanced yield from unit area, higher use efficiency of resources (soil nutrients, soil moisture, atmospheric CO_2, sunlight and land), conservation of resources and soil health management and enhancement of soil fertility (by checking erosion and nutrient loss with run-off of water) and agricultural sustainability (diversification, legume effect, less chemical use and ecosystem services). In other words, it may be stated that intercropping has enough potential for poverty alleviation, reduction of hunger, provisioning of healthy foods and biodiversity enhancement covering directly or indirectly some SDGs, such as "SDG 2 (end hunger, achieve food security and improved nutrition and promote sustainable agriculture), SDG 13 (take urgent action to combat climate change and its impacts) and SDG 15 (protect, restore and promote sustainable use of terrestrial ecosystems, sustainably manage forests, combat desertification, and halt and reverse land degradation and halt biodiversity loss)" (UN, 2021).

As in the intercropping system, more ground area is covered by the canopy of crop mixtures, the foliage does more transpiration resulting into the formation of a soothing microclimate and this helps to reduce the soil temperature (Innis, 1997; Miao *et al.*, 2016). Under water stress conditions, various crops in mixed stands use more available soil moisture and this microclimate creates a favourable condition facilitating congenial environment for crop growth (Mao *et al.*, 2012). In intercropping, preferably dissimilar crops based on morphology are grown and therefore, available and limiting resources are better used and as a result the combined biomass yield from a unit area increases (Lithourgidis *et al.*, 2011b). But there are various factors responsible for the success and failure of an intercropping system and these are crop choice, choice of cultivars (suitable for intercropping showing maximum complement in mixed stands), planting geometry, seeding proportion, and crop management inclusive of nutrients and water and plant protection from abiotic agents.

1.2 Types of Intercropping

Intercropping is the raising of more than one crop simultaneously on the same piece of land. In intercropping, crops are intensified spatially and temporally with various combinations of annual, biennial and perennial crops and the crops are selected on the basis of the farmers' choice and suitability of crops to the agroclimatic region and cultivation conditions (Eskandari *et al.*, 2009). The component crop species chosen in an intercropping competes for available resources in a mixture during a certain period or entire period of coexistence. Based on the spatial and temporal arrangements, intercropping systems adopted in different parts of the world can be categorized into the following groups (Ofori and Stern, 1987).

1.2.1 Row Intercropping

In row intercropping, one or more crops are sown in regular rows and component crops are sown simultaneously either in rows or without any row arrangement. Majority of the improved agricultural practices throughout the world use row intercropping to optimize productivity and resource use efficiency (Varma *et al.*, 2017).

1.2.2 Mixed Intercropping

The mixed intercropping is a practice of growing two or more crops together without any distinct row arrangement. It is also termed as mixed cropping. The common example of mixed intercropping is observed in the pasture-based cropping system, grass and legume mixed intercropping and mixed cover cropping in temperate countries etc. In general, it is practiced to fulfill the diverse need of food and forage from scarce land area (Undie *et al*., 2012) and mixed stands of winter cover crops during off-season in temperate regions for soil quality improvement and production enhancement of the succeeding main or cash crops.

1.2.3 Strip-intercropping

The strip-intercropping is another kind of intercropping system where two or more crops are cultivated simultaneously in different strips in sloppy lands. Like in other intercropping systems, it also has the capability to enhance radiation use efficiency (Yang *et al*., 2015) and it is practiced in the sloppy lands having a high possibility of soil erosion. In the agroforestry system, strip intercropping is commonly practiced. Generally, soil draining and conserving crops species are grown alternatively in the strips running perpendicular to the slope of the land or to the direction of prevailing winds for checking the soil erosion.

1.2.4 Relay Intercropping

In the relay intercropping, two or more crops are grown together and they co-exist for a portion of the growing period of each, but not sown at the same time. Here, the second crop is sown when the first crop is about to complete the vital part of its cycle or completes a major portion and reaches the reproductive stage or close to maturity but before harvest (Maitra *et al*., 2021). There are some areas where cropping season is a limiting factor for growing two crops sequentially and hence, relay intercropping is adopted to enhance the cropping intensity (Balde *et al*., 2011). Before completion of the life cycle of the preceding crop (standing situation), the second crop is sown and the second crop starts its cycle at optimum time as per the growing season. In this case, the first crop is harvested at maturity and the second crop gets the entire land to flourish during the remaining part of the cropping system, when essential care and management are provided to the second crop. The areas where double cropping is commonly adopted but due to climatic aberrations or other issues, harvest of the preceding crop is delayed, relay intercropping is the most suitable option for the situation for reaping two crops within the calendar year (Thiessen Martens *et al.,* 2005). Moreover, in rice-based cropping system, growing pulses as relay intercropping is very common in India and neighboring countries of South Asia where rice is harvested during end November or mid-December and growing the winter crops after preparatory tillage becomes difficult. Therefore, farmers adopt relay intercropping under that particular situation. Not only the pulses but also oilseeds and some other low input demanding crops are also considered under the situation. In the relay intercropping, seed rate of the second crop requires more for an optimum plant stand and generally low yields are obtained in field crops in comparison to the normal sowing practice adopted for the second crop in the sequential cropping. Also, in vegetables-based cropping system where *pondals* (temporary structures) are prepared to provide support to the vegetables with creeping habit, the succeeding crops are seeded before harvest of the preceding crop targeting an enhanced cropping intensity.

1.3 Crop Geometry and Types of Intercropping

The advantages and disadvantages of intercropping systems greatly depend on the ratio of component crops and crop geometry in order to create an efficient spatial arrangement compared to their individual sole stands (Yang *et al.*, 2015). Based on the crop geometry followed, intercropping can be classified into the following two groups.

1.3.1 Additive Series

In the additive series of intercropping, a crop is grown with 100% population of its pure stand and component crop or crops are within it on the same field at the same time. The crop grown with 100 per cent population is called main or base crop and the component crops are termed as intercrops. In this type of intercropping system, main crops are chosen with wider row spacing as intercrops can be accommodated within the space available between two rows of the main crop. Further, there is an option of modification of planting geometry for creation of more space to accommodate intercrops. The paired row planting is a method of creation of more space between two pairs of base crop. Maize, cotton, sugarcane, red gram are some crops where paired row planning method is commonly adopted (Figures 1.1, 1.2). In additive series of intercropping, either in uniform row or paired row planting, the base crop population remains optimum (i.e., 100) like sole cropping and the population of intercrops remain mostly less than 100% (Figure 1.3). In other words, it may be stated that additive series of intercropping is an addition of some population intercrops/ component crops in optimum stand of base crop. In most of the cases, if the intercrops are chosen less aggressive with a greater extent of complement, the land equivalent ratio (LER) is generally recorded as greater than unity (1.0) that indicates yield advantage of intercropping systems adopted. As the yield advantage is synonymous to additional gross return, additive series is recognized as an efficient intercropping system being preferred by the small farmers.

Figure 1.1. Additive series of intercropping groundnut in paired row maize in south Odisha

1.3.2 Replacement Series

In the replacement series of intercropping, no single crop gets its 100% population as it is maintained in the pure stand. In this intercropping system, the crop species cultivated simultaneously are termed as component crops or intercrops and a component crop is introduced by replacing another component crop (Figure 1.4). In the replacement series of intercropping, inter-species competition is less compared to additive series and yield advantage is obtained mainly due to a higher level of complement among the crop species grown. But to ensure yield advantage, sometimes seeding density of component crops is increased than the recommended plant population maintained in their sole stands (Baker and Blamey, 1985).

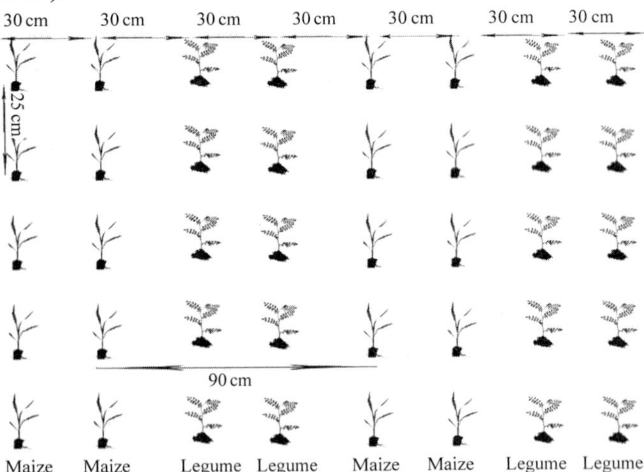

Figure 1.2. Paired row maize planted at a spacing of 30 cm/ 90 cm × 25 cm in which 2 rows of legumes incorporated in an additive series

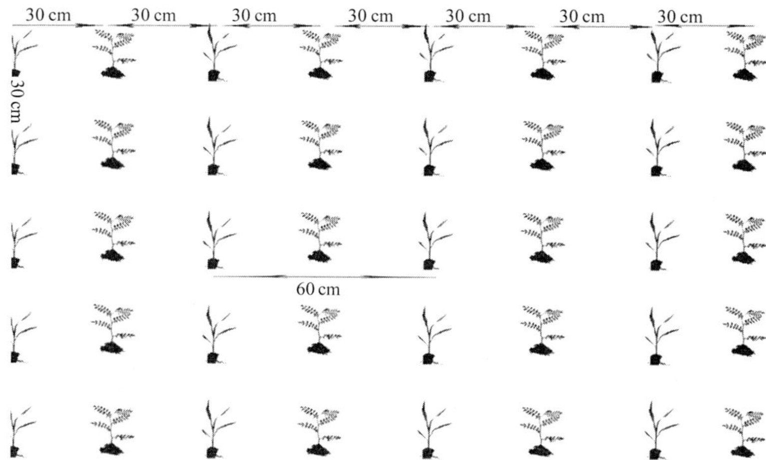

Figure 1.3. Uniform row maize planted at a spacing of 60 cm × 30 cm in which 1 row of legume incorporated in an additive series

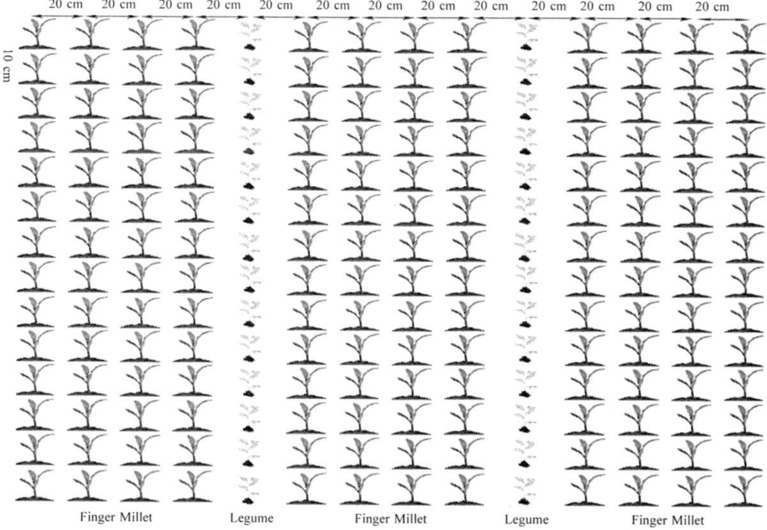

Figure 1.4. In a replacement series, 4 rows of finger millet sown at a spacing of 20 cm x 10 cm followed by 1 row of legume

1.4 Choice of Crops

The choice of crops is a vital consideration in relation to the growing conditions, crop season and agroclimatic conditions, local demand, marketability, availability of seeds and competitiveness of the crop species. Crops grown in mixed stands with higher levels of complement exhibit yield benefits. A study revealed that an intercropping combination of finger millet and pigeon pea or peanut (4:1) showed maximum net return, LER and other competitive functions; whereas in the same study, effect of finger millet + soybean or groundnut (4:1) was not prominent (Maitra *et al*., 2000). Similarly, maize + *faba* bean yielded more than wheat + *faba* bean (Fan *et al*., 2006). An experiment conducted in south Odisha clearly indicated that maize + groundnut (2:2) registered more LER, relative crowding co-efficient (RCC), area time equivalent ratio (ATER), relative yield total (RYT), maize equivalent yield and monetary advantage than the same proportion of maize intercropped with green gram and black gram (Manasa *et al*., 2018; 2019). Another important point is to be kept in mind before choosing the crops for intercropping. They should have physiological and morphological differences. Further, a combination of different crop duration is also preferred. All these features ensure greater use of limited resources and complement. The crops chosen for intercropping may be annual and perennial and their mixture. In alley cropping, a combination of perennials and annuals is observed where perennials form hedgerows and annuals are grown in alleys. Further, peak demand for resources of the crops should also be different in mixed stands as all the crop species chosen in the intercropping can utilize the resources. To make intercropping system more viable, breeding of suitable crop varieties is also been taken into consideration by using the latest breeding and simulation tools such as genomics, stochastic simulation, and so on (Bancic *et al*., 2021; Haug *et al*., 2021).

1.5 Intercropping Systems and Yield Advantages

Yield advantage is the basic aim of intercropping and intercropping systems may have to satisfy the requirements if yield advantages are to be achieved. For evaluation of the yield advantages of intercropping, the following situations can be notable (Willey, 1979).

a) *Where intercrop yield must give full yield of a base crop and some yield of a second crop.* Commonly, this practice is adopted in developing countries like India. In additive series generally full yield is obtained from base crop and some additional yield is achieved from intercrop.

b) *Where combined intercrop yield must exceed the higher sole crop yield.* In grassland mixture and cover cropping, generally the combined biomass yield exceeds their pure stands (Donald, 1963). In temperate regions, grasses and legumes are mixed to get higher combined yield that exceeds the maximum potential yield of the different crops grown separately.

c) *Where intercropping system yield must exceed summation of sole crop yields.* In this situation, yield advantages occur when maximum complement among the species is observed and intercropping results in more yield than pure stands of crops.

1.6 Limitations of Intercropping and Possible Solutions

There is no doubt that intercropping exhibits multifaceted advantages. But some limitations are also noted in intercropping over mono-cropping. Among different limitations, inter-species competition is the most important factor reducing yield of an intercropping system. Another vital concern is allelopathy that makes agronomic management difficult in mixed stands. The solution of these two limitations is crop choice and the focus can be given while choosing crops that they ensure a maximum level of complementary benefits and mutual sharing of limiting resources. If appropriate crop species are not chosen, there will be a high chance for reflection of disadvantages of intercropping in terms of reduced yield (Cenpukdee and Fukai, 1992; Santalla *et al*., 2001). Therefore, choice of crops and varieties, suitable proportions and planting geometry, and sowing time are vital to reduce competition among the species as well as limitations of intercropping.

The allelo-chemicals are bioactive chemicals released to surrounding soils by the leaves, stems, roots and decomposed plants and interact with the environment. If the interaction is negative, the allelo-chemicals produced from one species may show harmful effects on another plant. In different agro-forestry systems such as alley cropping and silvi-pastoral systems, black walnut (*Juglans nigra* L.) is a very common tree species and the allelo-chemical produced from it (juglone) shows harmful effects on companion species in mixed stands (Jose and Holzmuller, 2008). The leaf exudates of *Jatropha curcas*, a biofuel plant, exhibit negative impacts on seed germination and seedling vigour of chilli (Rejila and Vijayakumar, 2011). It can be noted that allelopathy is not always harmful and in multiple cropping system (inclusive of intercropping) allelo-chemical released by field crops suppress the weed growth and enhance crop yields in mixed stands (Narwal, 2000).

Another disadvantage of the intercropping system is difficulty in crop management practices where there is little or no scope for adoption of farm mechanization. Further, the component crops chosen in an intercropping system are preferably dissimilar and hence, they need different types of nutrients, water and other intercultural operations. The harvest and post-harvest operations are also difficult. There will be a chance of mixing grains of two or more crops grown in intercropping. Sometimes, at the time of harvest of early maturing crop in mixed stands, the late maturing crop may be mechanically damaged. To overcome these limitations, more labourers are required and in case of small farmers, underemployed family labourers can handle the harvest and post-harvest operations. But in the case of forage crops, such a problem does not occur (Lithourgidis *et al*., 2011a). Moreover, in some intercropping systems, the base crop or main crop yield may reduce compared to pure stands (Willey, 1979) because of inter-species competition for limiting resources. If the main crop yield is economically more meaningful, then it will be an economic loss. To manage the situation, intercrops with more economic value than base crop can be chosen for monetary advantage. There will be the chances of pest-disease incidence in intercropping system due to creation of microclimate and more humidity (Gliessman, 1985). But the microclimate is beneficial under limited soil moisture conditions and the functional diversity created in intercropping is also known to reduce the population of pests and pathogen (Maitra and Ray, 2019).

1.7 Annidation

The complementary interaction among component crop species occurring in intercropping is known as annidation. In intercropping, interaction among crops is commonly observed and under the mixed stands, the crop responses are altered. In most of the cases, one component species influences on the other (s). Annidation occurs both spatially and temporally.

1.7.1 Spatial Annidation (annidation in space)

Here, the complimentary interactions between the intercrops occur with regard to spatial position (space/place). The multistoried cropping is a prominent example of spatial annidation. The component species occupy different vertical layers or heights by spread-ing out their canopies or roots either in air or in soil. As far as aerial annidation is concerned, the taller intercrops occupy higher vertical layers and shorter intercrops the lower layers. Such taller component species are comparatively tolerant to strong light and high evaporative demand than shorter one. On the other hand, shorter component species are relatively shade-loving and acclimatized to high relative humidity. A suitable example of multistoried cropping is growing coconut + papaya + pineapple. The principle of spatial annidation may also occur in soil and crops with different rooting pattern uptake essential nutrients from different soil depths and horizons. Therefore, for efficient utilization of soil resources, intercrops of different root systems, viz., deep-rooted species + shallow rooted species are grown. For example, intercropping of shallow rooted cereals and legumes with deep root system is a very common practice.

1.7.2 Temporal Annidation (annidation in time)

The complementary interaction between the intercrops in the time as-pect is called temporal annidation. Such intercrops have different natural habit and zero competition. Both the component species have widely varying duration and different peak demands for light, soil moisture and essential nutrients, therefore com-petition amongst the component species is reduced. In the maize and green gram or black gram intercropping system, the peak nutrient demand for maize is at knee-height stage (i.e., six to seven weeks after sowing); whereas, it is around four to five weeks for green gram or black gram. Other intercrop-ping systems utilizing this principle are groundnut + pigeon pea, sorghum + pigeon pea, and so on. Actually, temporal annidation may occur when the period of maximum resource exploitation (nutrient, light, water etc.) may be continued for few weeks; but the species chosen for intercropping should be of different period for their maximum resource demand. For example, more Land Equivalent Ratios (LERs) were noted when a short duration cereal was considered for intercropping intercropped with long duration legume, red gram or other slow-growing, long-season crops (Rao and Willey, 1980).

In general, through adoption of efficient intercropping systems, farmers can achieve a close to the full productivity of the base crop with an additional yield from component crops by increasing population density of the intercrops. Hence, intercropping has enough potential to enhance income of smallholders in the areas where labour scarcity is not a problem. In intercropping, it is mandatory to mandatory to choose the right crops and cultivars with appropriate agronomic practices to obtain desired yield advantage and other benefits for agricultural sustainability.

References

Altieri, M. A. 1999. The ecological role of biodiversity in agro-ecosystems. *Agr. Ecosyst. Environ.* **74**:19-31.

Andrews, D. J. and Kassam, A. H. 1976. The importance of multiple cropping in increasing world food supplies. *In*: Papendick, R. I., Sanchez, P. A., Triplett, G. B. (*Eds.*) Multiple Cropping. ASA Special Publication 27, American Science of Agronomy, Madison, WI, USA.

Anil, L., Park. J., Phipps, R. H. and Miller, F. A. 1998. Temperate intercropping of cereals for forage: A review of the potential for growth and utilization with particular reference to the UK. *Grass Forage Sci.* **53**:301-317.

Baker, C. M. and Blamey, F. P. C. 1985. Nitrogen fertilizer effects on yield and nitrogen uptake of sorghum and soybean, grown in sole cropping and intercropping systems. *Field Crop Res.* **12**:233–240.

Balde, A. B., Scope, l. E., Affholder, F., Corbeels, M., Da Silva, F. A. M., Xavier, J. H. V. and Wery, J. 2011. Agronomic performance of no-tillage relay intercropping with maize under smallholder conditions in Central Brazil. *Field Crop Res*. **124**(2):240–251.

Bancic, J., Werner C. R., Gaynor R. C., Gorjanc, G., Odeny, D. A., Ojulong, H. F., Dawson, I. K., Hoad, S. P. and Hickey, J. M. 2021. Modeling illustrates that genomic selection provides new opportunities for intercrop breeding. *Front. Plant Sci.* **12**:605172. doi: 10.3389/fpls.2021.605172.

Bedoussac, L., Journet, E. P., Hauggaard-Nielsen, H., Naudin, C., Corre-Hellou, G., Jensen, E.S., Prieur, L. and Justes, E. 2015. Ecological principles underlying the increase of productivity achieved by cereal-grain legume intercrops in organic farming: a review. *Agron. Sustain. Dev.* **35**:911–935.

Bulson, H. A. J., Snaydon, R. W. and Stopes, C. E. 1997. Effects of plant density on intercropped wheat and filed beans in an organic farming system. *J. Agric. Sci.* **128**:59-71.

Cenpukdee, U. and Fukai, S. 1992. Cassava/legume intercropping with contrasting cassava cultivars. 1. Competition between component crops under three intercropping conditions. *Field Crops Res.* **29**:113-133.

Donald, C. M. 1963. Competition among crop and pasture plants. *Adv. Agron.* **15**:1-18.

Eskandari, H., Ghanbari, A. and Javanmard, A. 2009. Intercropping of cereals and legumes for forage production. *Nat. Sci. Biol.* **1**:07-13.

Fan, F., Zhang, F., Song Y., Sun, J., Bao, X., Guo, T. and Li, L. 2006. Nitrogen fixation of faba bean (*Vicia faba* L.) interacting with a non-legume in two contrasting intercropping systems. *Plant Soil.* **283**: 275-286.

Fininsa, C. and Yuen, J. 2001. Association of bean rust and common bacterial blight epidemics with cropping systems in Hararghe highlands, eastern Ethiopia. *Int. J Pest Manage.* **47**: 211–219.

Francis, C. A. 1986. Introduction: distribution and importance of multiple cropping. Pages 1-20, *In*: Multiple cropping systems (Francis, C.A., *Eds*.). Macmillan Publishing Company, New York, USA.

Fuller, D.Q. 2011. Pathways to Asian civilizations: Tracing the origins and spread of rice and rice cultures. *Rice.* **4**:78–92.

Gliessman, S. R. 1985. Agro-Ecological Processes in Sustainable Agriculture. Sleeping Bear Press, Chelsea, ML, USA.

Haug, B., Messmer, M. M., Enjalbert, J., Goldringer, I., Forst, E., Flutre, T., Mary-Huard, T. and Hohmann., P. 2021. Advances in breeding for mixed cropping – incomplete factorials and the producer/associate concept. *Front. Plant Sci.* **11**:620400. doi: 10.3389/fpls.2020.620400.

Innis, D. Q. 1997. Intercropping and the scientific basis of traditional agriculture. Intermediate Technology Publications Ltd, London.

Jackson, L. E., Pascual, U. and Hodgkin, T. 2007. Utilizing and conserving agro-biodiversity in agricultural landscapes. *Agr. Ecosyst. Environ.* **121**:196-210.

Jensen, E. S., Hauggaard-Nielsen, H., Kinane, J., Andersen, M. K. and Jornsgaard, B. 2005. Intercropping – The practical application of diversity, competition and facilitation in arable organic cropping systems. *In*: Kopke *et al.* (*eds*) Researching Sustainable Systems 2005. Proceedings of the First Scientific Conference of the International Society of Organic Agricultural Research (ISOFAR), Bonn, Germany, 2005, pp. 22-25.

Jose, S. and Holzmueller, E. 2008. Black Walnut Allelopathy: Implications for Intercropping. In: Zeng R.S., Mallik A.U., Luo S.M. (*Eds.*) Allelopathy in Sustainable Agriculture and Forestry. Springer, New York, NY.

Knörzer, H., Graeff-Hönninger, S., Guo, B., Wang, P. and Claupein, W. 2009. The Rediscovery of Intercropping in China: A Traditional Cropping System for Future Chinese Agriculture – A Review. In: E. Lichtfouse (ed.), Climate Change, Intercropping, Pest Control and Beneficial Microorganisms, Sustainable Agriculture Reviews 2, Springer Science + Business Media B.V., DOI 10.1007/978-90-481-2716-0_3, pp.13-44.

Lal, B. B. 1970. Perhaps the earliest ploughed field so far excavated in the world, *Puratattva*, **4**: 1-6.

Li, L., Li, S. M., Sun, J. H., Zhou, L. L., Bao, X. G., Zhang, H. G. and Zhang, F. 2007. Diversity enhances agricultural productivity via rhizosphere phosphorus facilitation on phosphorus deficient soils. *PNAS.* **104**(27), 11192–11196.

Li, L., Sun, J., Zhang, F., Guo, T., Bao, X., Smith, F. A. & Smith, S. E. 2006. Root distribution and interactions between intercropped species. *Oecologia.* **147**:280-290.

Li, Y., Ran, W., Zhang, R., Sun, S., Xu, G., 2009. Facilitated legume nodulation, phosphate uptake and nitrogen transfer by arbuscular inoculation in an upland rice and mung bean intercropping system. *Plant Soil.* **315**:285–296. https://doi.org/10.1007/ s11104-008-9751-9.

Lichtfouse, E., Navarrete, M., Debaeke, P., Souchere, V., Alberola, C. and Menassieu, J. 2009. Agronomy for sustainable agriculture. A review. *Agron. Sustain. Dev.* **29**:1-6.

Lithourgidis, A. S., Dordas, C. A., Damalas, C. A. and Vlachostergios, D. N. 2011a. Annual intercrops: an alternative pathway for sustainable agriculture. *Aust. J. Crop Sci.* **5**(4):396-410.

Lithourgidis, A. S., Vlachostergios, D. N., Dordas, C. A. and Damalas, C. A. 2011b. Dry matter yield, nitrogen content, and competition in pea–cereal intercropping systems. *Eur. J. Agron.* **34**:287–294.

Maitra, S. and Gitari, H. I. 2020. Scope for adoption of intercropping system in organic agriculture. *Ind. J. Nat. Sci.* **11**(63): 28624-28631.

Maitra, S. and Ray, D. P. 2019. Enrichment of biodiversity, influence in microbial population dynamics of soil and nutrient utilization in cereal-legume intercropping systems: A Review. *Int. J. Biores. Sci.* **6**:11–19, doi:10.30954/2347- 9655.01.2019.3.

Maitra, S., Barik, A., Samui, S. K. and Saha, D. 1999. Economics of cotton based intercropping system in the rice fallows of coastal Bengal- *Sundarbans. J. Ind. Soc. Coastal Agric. Res.* **17**(1-2): 299-304.

Maitra, S., Ghosh, D. C., Sounda, G. Jana, P. K. and Roy, D. K. 2000. Productivity, competition and economics of intercropping legumes in finger millet (*Eleusine coracana*) at different fertility levels. *Ind. J. Agric. Sci.* **70**(12): 824-828.

Maitra, S., Hossain, A., Brestic, M., Skalicky, M., Ondrisik, P., Gitari, H., Brahmachari, K., Shankar, T., Bhadra, P., Palai, J.B., Jena, J., Bhattacharya, U., Duvvada, S.K., Lalichetti, S., Sairam, M. 2021. Intercropping- a low input agricultural strategy for food and environmental security. *Agronomy.* **11**:343; https://doi.org/10.3390/ agronomy11020343.

Maitra, S., Palai, J. B., Manasa, P. and Kumar, D. P. 2019. Potential of intercropping system in sustaining crop productivity. *Int. J. Agric. Environ. Biotech.* **12**:39–45, doi:10.30954/0974–1712.03.2019.7.

Maitra, S., Shankar, T. and Banerjee, P. 2020. Potential and Advantages of Maize-Legume Intercropping System (Online First), Intech Open, DOI: 10.5772/intechopen.91722. Available from: https://www.intechopen.com (Accessed on 26 March, 2020).

Manasa, P., Maitra, S. and Barman, S. 2020. Yield attributes, yield, competitive ability and economics of summer maize-legume intercropping system. *Int. J. Agric. Environ. Biotech.* **13**(1): 33-38, DOI: 10.30954/0974-1712.1.2020.16.

Manasa, P., Maitra, S. and Devender Reddy, M. 2018. Effect of summer maize-legume intercropping system on growth, productivity and competitive ability of crops, *Int. J. Manage. Technol. Engi.* **8**(12): 2871-2875.

Mao, L., Zhang. L., Li, W., van der Werf, W., Sun, J., Spiertz, H. and Li, L. 2012. Yield advantage and water saving in maize/pea intercrop. *Field Crop Res.* **138**:11–20.

Miao, Q., Rosa, R. D., Shi, H., Paredes, P., Zhu, L., Dai, J., Gonçalves, J. M. and Pereira, L. S. 2016. Modeling water use, transpiration and soil evaporation of spring wheat–maize and spring wheat–sunflower relay intercropping using the dual crop coefficient approach. *Agric Water Manag.* **165**:211–229.

Mucheru-Muna M, Pypers P, Mugendi D, et al. (2010) A staggered maize-legume intercrop arrangement robustly increases crop yields and economic returns in the highlands of Central Kenya. *Field Crops Res.* **115**: 132–139.

Narwal S. S. 2000. Allelopathic interactions in multiple cropping systems. In: Narwal S.S., Hoagland R.E., Dilday R.H., Reigosa M.J. (eds.) Allelopathy in Ecological Agriculture and Forestry. Springer, Dordrecht. https://doi.org/10.1007/978-94-011-4173-4_10.

Ofori, F. and Stern, W.R. 1987. Cereal-legume intercropping system, *Adv. Agron.* **41**: 41-90.

Papanastasis, V. P., Arianoutsou, M. and Lyrintzis, G. 2004. Management of biotic resources in ancient Greece. Proceedings of the 10th Mediterranean Ecosystems (MEDECOS) Conference, 25 April-01 May 2004, Rhodes, Greece, pp. 1-11.

Plucknett, D. L. and Smith, N. J. H. 1986. Historical perspectives on multiple cropping. *In:* Francis, C. A., (*Eds.*) Multiple Cropping Systems. MacMillan Publishing Company. New York, USA.

Rao, M. R. and Willey, R. W. 1980. Evaluation of yield stability in intercropping: Studies on sorghum/pigeonpea. *Exp. Agr*. **16**:105-116.

Rejila, S. and Vijayakumar, N. 2011. Allelopathic effect of *Jatropha curcas* on selected intercropping plants (green chilli and sesame). *J. Phytol.* **3**(5):1-3.

Rodriguez, C., Carlsson, G., Englund, J-E., Flöhr, A., Pelzer, E., Jeuffroy, M. H., Makowski, D. and Erik Steen Jensen, E. S. 2020. Grain legume-cereal intercropping enhances the use of soil-derived and biologically fixed nitrogen in temperate agroecosystems. A meta-analysis. *Eur. J. Agron.* **118**:12607. https://doi.org/10.1016/j.eja.2020.126077.

Santalla, M., Rodino, A. P., Casquero, P. A., de Ron, A. M. 2001. Interactions of bush bean intercropped with field and sweet maize. *Eur. J. Agron*. **15**:185-196.

Scherr, S. J. and McNeely, J. A. 2008. Biodiversity conservation and agricultural sustainability: towards a new paradigm of 'ecoagriculture' landscapes. *Philos. Trans. Royal Soc. B*. **363**:477-494.

Thiessen Martens, J. R., Entz, M. H. and Hoeppner, J. W. 2005. Legume cover crops with winter cereals in southern Manitoba: fertilizer replacement values for oat. *Can. J. Plant Sci.* **85**(3):645–648.

Tilman, D., Cassman, K. G., Matson, P. A., Naylor, R. and Polasky, S. 2002. Agricultural sustainability and intensive production practices. *Nature*. **418**:671-677.

Tripathy, K. K. 2019. Budget 2019. Agriculture and Rural Prosperity, *Kurukshetra*. **67**(10): 5-8.

UN. 2021. Sustainable Development, The 17 Goals, Department of Economic and Social Affairs, https://sdgs.un.org/goals (Accessed on 11 November, 2021).

Undie, U. L., Uwah, D. F. and Attoe, E. E. 2012. Effect of intercropping and crop arrangement on yield and productivity of late season maize/soybean mixtures in the humid environment of south southern Nigeria. *J. Agric. Res*. **4**(4):37.

Vandermeer, J. H. 1989. The Ecology of Intercropping; Cambridge University Press: Cambridge, UK, 1989.

Varma, D., Meena, R. S. and Kumar, S. 2017. Response of mungbean to fertility and lime levels under soil acidity in an alley cropping system in Vindhyan region, India. *Int. J. Chem. Stu.* **5**(2):384–389.

Willey, R. W. 1979. Intercropping— Its importance and research Part-2. Agronomy and research approaches. *Field Crops Abst.* **32**:73-81.

Willey, R. W., and Reddy, M. S. 1981. A field technique for separating above-and below- ground interactions in intercropping: an experiment with pearl millet/groundnut. *Exp. Agric.* **17**:257-264.

Willey, R. W., Natarajan, M., Reddy, M. S., Rao, M. R., Nambiar, P. T. C., Kannaiyan, J. and Bhatnagar, V. S. 1983. Intercropping studies with annual crops. *In:* Better crops for food. Pitman, London, U.K., 83-100 pp.

Wright, R. P. 2010. The Ancient Indus: Urbanism, Economy, and Society: Case Studies in Early Societies; Cambridge University Press: New York, NY, USA, 2010.

Yang, F., Wang, X. C., Liao, D. P., Lu, F. Z., Gao, R. C., Liu, W. G., Yong, T., Wu, X., Du, J., Liu, J. and Yang, W. 2015. Yield response to different planting geometries in maize-soybean relay strip intercropping systems. *Agron. J.* **107**(1):296–304.

2

Competitive Relations and Functions

The competitive relations among the cultivated crop species perform a relation and these are pronounced during the early growth period of the crops and during later phases. The crop's performance is reflected into the yield output. Intercropping research so far conducted is based on the experimental layout of either replacement or additive series. The additive series of intercropping is suitable for the widely spaced and moderately tall statured crops; whereas, for the crops with close spacing the replacement series of intercropping is adopted. In the replacement series of intercropping, proportionate replacement is done by inclusion of intercrops. The additive series of intercropping is expected to exhibit benefits because preferably short duration crops are selected that completes their cycle before initiation of maximum demand for resources by the base crop. However, in the replacement series of intercropping, component crops are selected based on anticipated advantages over sole cropping of the crop species. Hence, it is of prime interest to evaluate the advantages of the intercropping system.

The main focus of the intercropping system is to understand the underlying mechanism to design an "optimal" polyculture as the crop species coexist can flourish appropriately by expressing their yield. In this regard, various crop densities and row proportions are adopted to ascertain maximum mutual and complementary effects among the crop species. In most of the studies, yield and income benefits are measured to evaluate the efficiency of mixed stands. There is no doubt that yield enhancement is the ultimate goal. But crop intensification and diversification should also be considered in the present context where achieving agricultural sustainability is a major concern for both the developed and developing countries.

Based on the primary goal for yield enhancement, the chapter focuses on important competitive relations among the crop species grown in mixed stands.

2.1 Competitive Relations

In an intercropping system, at least two crops are grown and these show competitions among themselves. Majority of researches conducted on intercropping were based on competitive relations between two species and on replacement series as well. In the replacement series, one crop species is replaced by the others with almost equivalent row proportions. The crops grown in the mixed stand show relationship among themselves in the following manners.

i) Competitive

In the competitive relationship, the yield of one crop increases through reduction of yield of another because of their respective population in the mixed stand. Such a relationship is also termed as 'compensation'. It indicates that yield reduction of one crop (because of replacement) may be compensated by another crop (which one is included). Willey (1979) denoted the two crops grown in this intercropping system as 'dominant' and 'dominated' crops.

ii) Complementary

In a complementary relationship, increased output of one species facilitates an enhanced output of the other crop and it is also known as 'mutual cooperation' (Willey, 1979). Generally, such relationships are rare.

iii) Supplementary

The supplementary relationship is observed when yield of one species increases without impacting on the yield of another crop species. A wide difference (physiologically and morphologically) between the crop species when chosen in an intercropping system, occurrence of supplementary relationship may be recorded.

iv) Mutual Inhibition

In case of the mutual inhibition, the yield output of each component species becomes less than the expected. Such type of intercropping is rarely practiced.

Worldwide, intercropping is practiced mainly in the developing and agriculture-based countries and in most of the cases, the competitive and supplementary relationships are prominently observed.

2.2 Competitive Functions

Majority of the research conducted on intercropping systems indicating the valuable assessment on differences between sole and intercropping were developed in the period from 1970 to 1980. In this regard, the land equivalent ratio (LER) was developed and later, the LER became the most commonly used expression to assess the efficiency of an intercropping system (Willey and Osiru, 1972; Willey, 1979; Beets, 1982; Spitters and van den Bergh, 1982). Some time after, researchers reviewed the relevance of the LER and they went with it (Francis, 1986; Ofori and Stern, 1987; Fukai 1993). One thing should be remembered here that most of the experiments conducted to assess the competitive ability were on replacement series comprised of two species. The studies focused on evaluating comparative yields between sole and intercropping of crops considered. In case of additive series of intercropping, the base crop population always remains 100% and additionally intercrops are added; hence, the LER value is exhibited as more than unity (Willey and Osiru, 1972; Harper, 1977; Mead and Willey, 1980). Therefore, scientists clinched that the LER value is the researchers' concern in evaluating the efficiency of intercropping over sole crops (Anders *et al.*, 1996).

Other than LER, researchers developed some more expressions to describe the competition among the crop species in the intercropping system. All the competitive functions developed so far are listed here and described.

2.2.1 Land Equivalent Ratio (LER)

The concept of land equivalent ratio (LER) was given by Willey and Osiru (1972). It expresses that the proportionate land area required under sole crop to produce the same yield as recorded in an intercropping system at the same level of management practices adopted (Willey *et al.*, 1983; Fetene, 2003). Actually, it is the summation of ratios of the yield of each crop considered in an intercropping system to its corresponding pure stand yield. The LER of each species is calculated separately and the LER values of each species are added to obtain a combined LER of the intercropping system. The LER of crop in pure stand is considered as unity.

In a replacement series of intercropping system with two crops, the LER is computed by the following formula when the crops proportion is 50:50.

$$\text{LER} = \frac{Yab}{Yaa} + \frac{Yba}{Ybb} = La + Lb$$

In the formula, Yab is the yield of "*a*" crop when intercropped with "*b*" crop and Yba is the yield of "*b*" crop cultivated in an intercropping system with "*a*" crop. Yaa and Ybb are the yields of "*a*" and "*b*" crops cultivated in pure stands, respectively.

The modified formula for any other situation (other than 50:50 population) is given below.

$$\text{LER} = \frac{Yab}{Yaa \text{ X } Zab} + \frac{Yba}{Ybb \text{ X } Zba}$$

The LER specifies the benefits of an intercropping system in utilizing the limited resources as against their pure stand (Mead and Willey, 1980). The value of LER greater than unity (1.0) indicates the advantages of the intercropping system (Ofori and Stern, 1987). In contrast, when LER is lower than one (1.0), the intercropping system is considered as disadvantageous (Caballero *et al.*, 1995). If the LER value of an intercropping system is 1.43, it means there is 43% yield advantage in intercropping compared to the cultivation of pure stands of the respective crops.

In the following tables (Table 2.1 and 2.2), the LER values of some field experiments on different intercropping systems carried out have been presented.

Table 2.1. The LER obtained in the experiments conducted on additive series of intercropping

Intercropping combination	Row ratio/ seeding proportion	LERa*	LERb*	Combined LER	References
Maize + common bean	1:1	1.11	0.49	1.60	Yilmaz *et al.*, 2008
Maize + common bean	2:2	1.09	0.39	1.48	
Maize + cowpea	1:1	1.10	0.58	1.68	
Maize + cowpea	2:2	1.11	0.48	1.59	
Maize + green gram	2:1	0.87	0.82	1.69	Manasa *et al.*, 2018
Maize + green gram	2:2	0.88	0.89	1.77	
Maize + soybean	100:50	0.89	0.31	1.20	El-Karamity *et al.*, 2020
Maize + groundnut	100:50	0.95	0.49	1.44	
Maize + sesame	100:50	0.85	0.45	1.30	
Maize + groundnut	2:1	0.89	0.86	1.75	Panda *et al.*, 2021
Maize + groundnut	2:2	0.92	0.78	1.70	
Maize + groundnut	2:3	0.92	0.82	1.74	

*LER*a* and LER*b* are the corresponding LER values of the first and second crop, respectively.

As mentioned earlier, in the additive series of intercropping, the base crop is sown with 100% population compared to its pure stand and additionally intercrops are sown. Therefore, because of 100% population density base crop yield is not reduced much; however, the yield of intercrop is an added advantage from the unit area. Hence, in additive series of intercropping, the LER values generally appear as more than 1.0 indicate yield advantage.

Table 2.2. The LER obtained in the experiments conducted on replacement series of intercropping

Intercropping combination	Row ratio/ seeding proportion	LER*a**	LER*b**	Combined LER	References
Finger millet + red gram	4:1	0.77	0.72	1.49	Maitra *et al.*, 2000
Finger millet + green gram	4:1	0.77	0.27	1.04	
Finger millet + groundnut	4:1	0.70	0.30	1.00	
Finger millet + soybean	4:1	0.79	0.28	1.07	
Groundnut + sesame	2:2	0.79	0.54	1.33	Khan *et al.*, 2017
Groundnut + sesame	3:2	0.87	0.77	1.64	

*LER*a* and LER*b* are the corresponding LER values of the first and second crop, respectively.

In the replacement series of intercropping, the crops are sown with less than 100% percent population and yield benefits are obtained due to complementary and supplementary relationship among the crop species. In the table 2.1, with the combination of finger millet + groundnut, the combined LER was 1.0 indicating neither advantage nor disadvantage of intercropping. However, the LER value of less than unity denotes the disadvantage of the intercropping system.

2.2.2 Area Time Equivalent Ratio (ATER)

The LER has a vital limitation as it accentuates on only land area, but time factor is not considered for which the crop remains in the field. Therefore, scientists were in search of another expression considering the field occupancy by the crops in an intercropping to precise the limitation of the LER. Hiebsch (1978) worked on it and expresses the concept of Area Time Equivalent Ratio (ATER) where the duration of crops and field occupancy by them were considered. The formula for calculation of the ATER is given below.

$$\text{ATER} = \frac{(\text{RY}c \text{ X t}c) + (\text{RY}p \text{ X t}p)}{\text{T}}$$

Where, RY = Relative yields of crop species "*c*" and "*p*"

$$\text{RY} = \frac{\text{Yiel of intercrop/ha}}{\text{Yiel of sole crop/ha}}$$

t = Duration (in days) for species "*c*" and "*p*"

T = Duration (in days) for the intercropping system

However, the LER is known to overestimate and the ATER to underestimate the land-use efficiency (Sarkar and Shit, 1993; Maitra *et al.*, 2000). In the following table (Table 2.3) some LER and ATER values are presented where ATER values are less because of considering the time factor.

Table 2.3. The LER and ATER values from earlier researches

Intercropping system	Row ratio	LER	ATER	References
Chickpea + linseed	1:1	0.96	0.83	Sarkar *et al.*, 2000
Chickpea + linseed	1:2	1.09	0.79	
Chickpea + linseed	2:1	1.07	0.80	
Chickpea + toria	1:1	1.14	0.67	
Chickpea + toria	1:2	1.13	0.62	
Chickpea + toria	2:1	1.14	0.73	
Maize + groundnut	2:1	1.75	1.06	Panda *et al.*, 2021
Maize + groundnut	2:2	1.70	1.26	
Maize + green gram	2:1	1.65	0.97	
Maize + green gram	2:2	1.62	1.07	

The above table clearly showed that the ATER value remained less than the LER for the same combination and proportion of crop species chosen for intercropping. Interestingly, where the duration of crops is close, the difference between the LER and ATER values is less. On the other hand, crops of dissimilar duration and maturity periods are preferably chosen in an intercropping system and the combination of short and long duration crops make a wider difference between the values of above competition functions. The combination with one short duration crop reduces the ATER. Sarkar *et al.* (2000) recorded narrower difference when chickpea intercropped with linseed than the chickpea + toria. Likewise, Panda *et al.* (2021) observed that the combination of maize + green gram yielded less ATER than maize + groundnut.

2.2.3 Aggressivity

The concept of aggressivity being proposed by McGilchrist (1965) indicates a modest measurement of the quantity of the relative yield enhanced in crop species "*a*" rather than crop species "*b*". Aggressivity is represented as "A". In a replacement series of experiment, aggressivity is computed by the following expression.

$$\text{Aab} = \frac{\text{Mixture yield of } a}{\text{Expected yield of } a} - \frac{\text{Mixture yield of } b}{\text{Expected yield of } b}$$

For other than replacement series of intercropping, the aggressivity is expressed by the formula given below.

$$\text{Aab} = \frac{Yab}{Yaa \times Zab} - \frac{Yba}{Ybb \times Zba}$$

$$\text{Aba} = \frac{Yba}{Ybb \times Zba} - \frac{Yab}{Yaa \times Zab}$$

Yab = Yield of crop "*a*" in intercropping

Yaa = Yield of crop "*a*" in sole cropping

Zab = Sown proportion of crop "a" in intercropping

Yba = Yield of crop "*b*" in intercropping

Ybb = Yield of crop "*b*" in sole cropping

Zba = Sown proportion of crop "*b*" in intercropping

When the aggressivity (A) value appears as zero, it denotes that both the crops are equally competitive and none of them are more aggressive. However, the positive (+) aggressivity value appears indicating "*a*" crop as aggressive or dominant species over "*b*". But, when the aggressivity value becomes negative (-), the "*b*" crop is recognized as aggressive or dominant over "*a*". In general, based on the competitive nature of the crops in the mixed stand, aggressivity value is expressed (Table 2.4).

Table 2.4. Aggressivity of some cereal- legume treatments studies in intercropping experiments

Intercropping system	Row ratio	A crop "a"	A crop "b"	
Cotton + green gram	2:1	0.31	-0.31	Maitra *et al*., (2001)
Cotton + green gram	1:2	0.26	-0.26	
Cotton + groundnut	2:1	0.50	-0.50	
Cotton + groundnut	1:2	0.11	-0.11	
Finger millet + red gram	4:1	-2.44	2.44	Maitra *et al*. (2000)
Finger millet + green gram	4:2	-0.46	0.46	
Maize + groundnut	2:1	0.06	-0.06	Manasa *et al*. (2018)
Maize + groundnut	2:2	0.13	-0.13	
Maize + green gram	2:1	0.06	-0.97	
Maize + green gram	2:2	0.12	-0.12	

In the cotton based intercropping system, cotton appeared as more aggressive crop as it registered its dominance over legumes, namely, green gram and groundnut Maitra *et al*., 2001. Similarly, Manasa *et al*. (2018) observed that maize was the dominant species. But, Maitra *et al*. (2000) recorded that finger millet showed the negative "A" values when intercropped with legumes in uplands of red and lateritic belt of West Bengal under rainfed conditions. Further, greater "A" value was recorded with the combination of finger millet and red gram indicating wider variation in competitive ability between crop species, resulting in a larger difference between the actual and expected yields of the crop species.

2.2.4 Relative Crowding Coefficient (RCC)

The Relative Crowding Coefficient (RCC) is a competition function which is indicative of relative dominance of one component species over another. The concept of RCC has been proposed by De Wit (1960) and examined in detail by Hall (1974a; 1974b). In an intercropping system, each crop has its own coefficient (K) that quantifies whether the species has yielded more or less than the expected yield. The following expression is used to calculate the relative crowding coefficient.

For species "*a*" in a 50:50 mixture with species "*b*", RCC ismeasured as:

Product of RCC (K) = K*ab* X K*ba*

K*ab* = (Y*ab*)/(Y*aa*-Y*ab*)

= (Mixture yield of *a*) / (Pure yield of *a* – Mixture yield of *a*)

Kba = (Y*ba*)/(Y*bb*-Y*ba*)

= (Mixture yield of *b*) / (Pure yield of *b* – Mixture yield of *b*)

Further, for a mixture differing from 50:50 proportion, RCC can be formulated as.

K*ab* = (Y*ab*XZ*ba*)/ (Y*aa*-Y*ab*) X Z*ab*

K*ba* = (Y*ba*X Z*ab*)/ (Y*bb*-Y*ba*) X Z*ba*

Where, K is the product of RCC

K*ab* and K*ba* are RCC for the crop species "a" and "b" respectively

Y*ab* = Yield ofcrop"*a*" in intercropping

Z*ba* = Sown proportion of crop "*b*" in intercropping

Y*aa* = Yield of crop "*a*" in sole cropping

Z*ab* = Sown proportion of "*a*" in intercropping

Y*ba* = Yield of crop "*b*" in intercropping

Y*bb* = Yield of legume "*b*" in sole cropping

By calculating the RCC, efficiency of intercropping is measured. The RCC value (i.e., product of two coefficients, K*ab* × K*ba*) computed more than one (>1) is indicative of benefits in the intercropping. Similarly, the RCC value of<1 indicates disadvantages of intercropping and if it appears as 1, it denotes neither advantage nor disadvantage of intercropping system. In the following table, RCC values have been extracted from the earlier research (Table 2.5).

Table 2.5. Relative crowding coefficient in cereal-legume intercropping system

Intercropping system	Row ratio	K_{Cereal}	K_{Legume}	Product of RCC (K)	References
Finger millet + red gram	4:1	4.67	10.75	50.20	Maitra *et al.*, 2000
Finger millet + green gram	4:1	2.92	1.89	5.52	
Finger millet + groundnut	4:1	4.18	2.14	8.94	
Finger millet + soybean	4:1	3.58	2.02	7.23	
Maize + cowpea	1:1	28.88	0.76	21.72	Choudhary, 2014
Maize + cowpea	1:2	41.69	2.15	89.63	
Maize + cowpea	1:5	1.50	4.36	6.54	

Maitra *et al.* (2000) conducted a study on finger millet and legume intercropping systems with replacement series and noted that the product (K) of RCC was more than 1 and it indicated yield advantages with all legume combinations. Similarly, Choudhary (2014) carried out an experiment in Arunachal Pradesh and recorded that the product (K) of RCC was greater than 1.0 with different proportions of maize + cowpea. The complementary effect between the crop species was responsible for yield advantage in intercropping.

2.2.5 Competitive Ratio (CR)

Competitive ratio (CR) is the competitive ability of the crops grown in combination in an intercropping (Willey and Rao, 1980). The CR indicated the number of times by which one crop species is more competitive than other (Willey and Rao, 1980) & it actually signifies the proportion of individual LERs of the crops considered in intercropping and also takes into account the ratio of the crops sown in a mixed stand. The CR can be expressed by the formulae given below.

$$CRa = (LERa/LERb) \times (Zba/Zab)$$

$$CRb = (LERb/LERa) \times (Zab/Zba)$$

Where, CR*a* and CR*b* are indicative of the competitive ratios of the crop species "*a*" and "*b*"and LER*a* and LER*b* are the LERs of the crop species "*a*" and "*b*" respectively. *Zab* is the sown ratio of species "*a*" in mixture with "*b*" and *Zba* is the sown proportion of the species "*b*" in mixture with "*a*". If the value of CR is < 1, there is a positive benefit and it means there is limited competition between component crops and therefore they can be grown as intercrops (Ghosh 2004). If the CR value is more than one (CR > 1), there is a negative impact. In this condition, the competition between intercrops in the mixture is too high, and they are not recommended to grow as intercrops. Further, Willey and Rao (1980) clearly mentioned, "the CR term is, therefore, simply the ratio of the individual

LERs of the two component crops, but correcting for the proportions in which the crops were initially sown". In the following table, data on aggressivity and competitive ratio have been extracted from the research of Choudhary (2014) and narrated (Table 2.6).

Intercropping system	Row ratio	A_{Maize}	A_{Legume}	CR_{Maize}	CR_{Legume}
Maize + cowpea	1:1	6 0.43	0.43	0.69	1.44
Maize + cowpea	1:2	6 0.13	0.13	0.88	1.13
Maize + cowpea	1:5	6 0.06	0.06	0.95	1.06
Maize + French bean	1:1	6 0.66	0.66	0.60	1.73
Maize + French bean	1:2	6 0.19	0.19	0.85	1.20
Maize + French bean	1:5	6 0.12	0.12	0.89	1.13
Maize + black gram	1:1	6 0.22	0.22	0.81	1.23
Maize + black gram	1:2	6 0.12	0.12	0.89	1.13
Maize + black gram	1:5	6 0.11	0.11	0.90	1.11

A_{Maize}= Aggressivity of maize; A_{Legume}= Aggressivity of legume; CR_{Maize}= Competitive ratio of maize; CR_{Legume}=Competitive ratio of legume

The above table (Table 2.6) clearly shows that the aggressivity values of maize with all combinations are negative and it indicates that maize is less aggressive than legumes. Similarly, the CR_{Maize} values are also less than one indicating legumes are more competitive. But, the CR_{Legume} values are just more than 1 and even less than 2 (ranged between 1.11 to 1.73). In the row proportion of 1:5, all legumes resulted in reduced CR values, indicating reduced inter-species competition.

2.2.6 Monetary Advantage

The concept of monetary advantage (MA) has been suggested by Willey (1979) and it is calculated by the expression mentioned below.

$$\text{Monetary advantage}\left(\text{MA}\right) = \frac{\text{LER} - 1}{\text{LER}} \text{ X value of combined intercrop yield}$$

The value of the produces is calculated on the basis of prevailing local market price. The monetary advantage is also referred as monetary advantage index (MAI). The higher value of the monetary advantage (MA) indicates more profitability of the intercrop combination (Ghosh, 2004).

2.2.7 Relative Yield Total (RYT)

The performance of different component crops in an intercropping system of replacement series was characterized by De Wit and van den Bergh (1965) and the concept of relative yield total (RYT) was generated. The RYT is the total of the relative yields of the species in an intercropping system and is expressed as the proportion of the yield of a crop species in the intercropping to its productivity in the pure stand (Anders *et al.*, 1996).The formula of RYT is as follows.

$$\text{RYT} = r_a + r_b + \text{----} + r_n$$

Where, r_a and r_b are the relative yields of species "*a*" and "*b*", respectively, computed as the proportion of intercropped and sole crop yields. The RYT values more than unity (1) denote that the both the crops are partially complementary; whereas, values less than one designate equal competitiveness of both the species. If the RYT value is less than one, it is better to grow the crop species separately.

2.2.8 Effective Land Equivalent Ratio (ELER)

As the different intercrop combinations contain dissimilarities of the crop species, the combinations with the largest LER will not necessarily be the superior as the productivity of one crop may not be sufficient. However, combinations of an intercrop and one of the sole crops can be used to ensure that the component crop yields are in the required proportion. The LER for this combination can then be calculated and this is the Effective Land Equivalent Ratio (ELER) as suggested by Mead and Willey (1980). The formula of ELER is as follows.

$$\text{ELER} = \frac{L_B}{(1-L_A) + (\text{LER}-1)p}$$

Where, $L_A = Y_A / S_A$ and $L_B = Y_B/S_B$

Y_A = Yield of *A* crop in intercropping

Y_B = Yield of *B* crop in intercropping

S_A = Yield of *A* crop grown in pure stand

S_B = Yield of *C* crop grown in pure stand

p = Required proportion of crop *A*, when sole crop of *A* is grown in addition to the intercrop

Further, Mead and Willey (1980) added, "these general equations assume that (a) the land area allocated to sole cropping would produce the same level of yield as the sole crop used in the L_A calculation, and (b) crop *A* is whichever of the two that has to be increased in proportion and of which there must therefore be some sole crop. The Effective LER is lower for any other proportion and, since the required proportion tends towards one or zero, the effective LER tends towards 1, indicating the use of either sole crop." If a pure stand of *B* is grown in addition to the intercrop, then L_A or L_B should be interchanged in the formula (Chatterjee *et al.*, 1989).

2.2.9 Relative Value Total (RVT)

In the LER, the combined yield values of the two crops are considered. Alternative method of combination could be based on their relative monetary values. For this purpose, Schultz *et al.* (1982) suggested the concept of Relative Value Total (RVT) for each intercropping. Following is formula of RVT.

$$\text{RVT} = (\text{Vi} + \text{Vii}) / \text{Vs}$$

Where, Vi and Vii are the monetary values of i and ii crops from intercrop treatments Vs is the appropriate sole crop monetary values

If the RVT value is more than 1, then intercrop treatment gives more return than pure stands. A disadvantage of RVT is that it does not take into consideration of the costs incurred in the production.

2.2.10 Relative Net Return (RNR)

Jain and Rao (1980) suggested an index to measure the efficiency of an intercropping system, the relative net return (RNR), in which comparative cost of cultivation was considered and different intercrops were compared with major crop. Following is the formula for calculation of RNR.

$$\text{RNR} = \left[(\text{PiYi} + \text{PjYj})\right] \pm \text{Dij}] / \text{PiYii}$$

Where, Y*i* and Y*j* are the yields of *i*th major crop and *j*th intercrop per hectare, respectively,

P*i* and P*j* are the prices of *i*th major crop and *j*th intercrop respectively, Y*ii*is the yield of *i*th sole crop, D*ij* is the differential cost of cultivation of *ij*th intercropping compared to sole crops.

2.2.11 Income Equivalent Ratio (IER)

Income Equivalent Ratio (IER) has the similarity to the concept of LER and here income is considered, rather than crop yields (kg/ha).

Ghaffarzadeh (1997) developed the formula for calculation of the IER.

$$\text{IER} = \frac{\text{Gross income / ha of intercropped crop A}}{\text{Gross income / ha of sole crop crop A}} + \frac{\text{Gross income / ha of intercropped crop B}}{\text{Gross income / ha of sole crop crop B}}$$

Like LER, the IER value when results more than unity, indicates advantage in intercropping. But the limitation of IER is the gross income is a variable factor and related to the market price. If crop yields remain same due to the variation in market price, income is changed. However, gross income includes byproducts also in some crops (for example, in rice farming, grain is the most economic part, straw is also sold and gross income is acquired). To calculate the IER, gross income (GI) obtained in intercropping from unit area is used.

2.1.12 Crop Equivalent Yield

In crop equivalent yield (CEY), the different crop yields are converted into one form for comparison (De Wit, 1960) and it is a common practice for evaluation of the efficiency of an intercropping or sequential cropping system. The yields of different crops are converted into base crop (A) equivalent yield by keeping into consideration of intercrop yields, and prevailing market price of 'A' crop, that is,base crop and intercrops. In simple words it can be stated that it is calculated by converting intercrop(s) yield into base crop yield and expressed in kg ha^{-1} based on local market price. The following formula is used to calculate the CEY.

$$\text{'A' Crop equivalent yield} = \frac{\text{Yield of intercrop x Price of intercrop}}{\text{Price of base (A) crop}} + \text{Yield of base (A) crop}$$

If the base crop equivalent yield is obtained higher in intercropping combinations than sole base crop yield, then intercropping is considered advantageous.

2.1.13 Actual Yield Loss (AYL)

The actual yield loss (AYL) is the proportionate yield loss or gain of component species in an intercropping system in compared to the respective pure stand. The AYL is based on the actual sown proportion of the component species in respect to its sole stand. Following is the formula for calculation of AYL for a cereal- legume intercropping system (Banik, 1996).

$\text{AYL} = \text{AYL}_{\text{cereal}} + \text{AYL}_{\text{legums}}$

Where, AYL cereal and AYL legume (partial actual yield loss of cereal and intercrop, respectively) represent the proportionate yield loss or gain of cereals and legumes in intercropping, about their yield in sole cropping. Further, they are calculated by the following formulae.

$\text{AYL}_{cereal} = \{\text{Yab / Zab}\} / \{(\text{Yaa/Zaa}) -1\}$

$\text{AYL}_{legume} = \{\text{Yba / Zba}\} / \{(\text{Ybb/Zbb}) -1\}$

Yab = Yield of cereal "a" in intercropping

Zab = Sown proportion of cereal "a" in intercropping

Yaa = Yield of cereal "a" in pure stand

Zaa = Sown proportion of cereal "a" in pure stand

Yba = Yield of legume "b" in intercropping

Zba = Sown proportion of legume "b" in intercropping

Ybb = Yield of legume "b" in pure stand

Zbb = Sown proportion of legume "b" in sole cropping

The AYL may have either positive or negative values indicating advantage or disadvantage in intercropping, respectively, when the yield is compared on per plant basis (Banik *et al.,* 2000). The AYL index gives precise information on inter and intraspecific competition in an intercropping system as well as the behavior of component crops (Banik *et al.,* 2000).

2.1.14 Intercropping Advantage (IA)

Intercropping advantage (IA) is calculated by comparing the advantage of intercropping in the monetary term (Banik *et al.,* 2000). In cereal-legume intercropping system, the formula of IA is expressed as follows.

$\text{IA}_{\text{Cereal}} = (\text{AYL}_{\text{Cereal}}) \text{ X } (\text{P}_{\text{Cereal}})$

$\text{IA}_{\text{Legums}} = (\text{AYL}_{\text{Legums}}) \text{ X } (\text{P}_{\text{Legums}})$

$\text{IA} = \text{IA}_{\text{Cereal}} + \text{IA}_{\text{Legums}}$

Where, P_{Cereal} is the commercial value of intercropped cereal and P_{Legume} is the commercial value of intercropped legume. If the partial IA values for intercropped cereals are positive, it means that intercrop cereals got a certain advantage due to intercropping with a legume. According to Banik *et al.* (2000), in addition to expressing the advantage or disadvantage of intercrops, IA can be an indicator of the economic feasibility of intercropping systems.

In the chapter, some important competitive functions have been presented to evaluate the efficiency of the intercropping system. As there is limited scope for enhancement of land horizontally for fulfillment of the need of the ever-growing human population, intensification can be made by adoption of an intercropping system where under limited land resource, gross crop productivity can be enhanced. Further, the intercropping system has enough potential in enhancing biodiversity, farmers' income and food as well as nutritional security, and agricultural sustainability. In the present circumstances of global warming and climate change, intercropping possesses versatile qualities for mitigation as well as reduction of carbon footprint in agriculture. The above-mentioned competitive functions will guide the researchers and policy makers to evaluate the scope and practical suitability for adoption of an intercropping system considering the present demand and future needs. Unless the efficiency of an intercropping system is properly evaluated, it neither be adopted in on a large scale nor advocated to adopt. Hence, the chapter will guide the concerned for the future endeavors towards achieving agricultural sustainability and maximizing yield output from the limited resources.

References

Anders, M. M., Potdar, M. V. and Francis, C. A. 1996. Significance of intercropping in cropping systems. *In:* Ito *et al.* (*Eds.*), Dynamics of roots and nitrogen in cropping systems of the Semi-Arid Tropics, Japan International Research Center tor Agricultural Sciences, ISBN:4-906635-01-6.

Banik, P. 1996. Evaluation of wheat (*Triticum aestivum*) and legume intercropping under 1:1 and 2:1 row replacement series system. *J. Agron. Crop Sci.* **176**:289–294.

Banik, P., Samsal, T., Ghosal, P. K. and Bagchi, D. K. 2000. Evaluation of mustard (*Brassica campestris* var. toria) and legume intercropping under 1:1 and 1:2 row replacement series system. *J. Agron. Crop Sci.* **185**:9–14.

Beets, W. C. 1982. Multiple cropping and tropical farming systems. Boulder, Colorado: Westview Press. 220p.

Caballero, R., Goicoechea, E. L. and Hernaiz, P. J. 1995. Forage yields and quality of common vetch and oat sown at varying seeding ratios and seeding rates of vetch. *Field Crop Res.* **41**:135–140.

Chatterjee, B. N., Maiti, S. and Mandal, B.K. 1989. Cropping System, Theory and Practices, Second Edition, Oxford and IBH Publishing Co. Pvt. Ltd., India.

Choudhary, V. K. 2014. Suitability of maize-legume intercrops with optimum row ratio in mid hills of Eastern Himalaya, India. *SAARC J. Agri.* **12**(2): 52-62.

De Wit, C. T. 1960. On competition. Verslag Land bouwkundige Onderzoekingen. *Wageningen.* **66**:1–81.

De Wit, C. T. and van den Bergh, J.P. 1965. Competition among herbage plants. *Neth. J. Agric. Sci.* **13**:212-221.

El-Karamity, A.E., Aya N. Mohamed, A. N. and Ahmed, N. R. 2020. Effect of Intercropping of Some Oil Summer Crops with Maize under Levels of Mineral N and Nano N Fertilizers. *Sci. J. Agric. Sci.***2**(2): 90-103, DOI: 10.21608/sjas.2020.42913.1038 90.

Fetene, M. 2003. Intra-and inter-specific competition between seedlings of *Acacia etbaica* and a perennial grass (*Hyparrhenia hirta*). *J. Arid Environ.* **55**:441–451.

Francis, C. A. 1986. Introduction: distribution and importance of multiple cropping. Pages 1-20, *In*: Multiple cropping systems (Francis, C.A., *Ed.*). Macmillan Publishing Company, New York, USA.

Fukai, S. 1993. Intercropping-basis of productivity. *Field Crops Res.* **34**:239-245.

Ghaffarzadeh, M. 1997. Economic and biological benefits of intercropping berseem clover with oat in corn-soybean oat rotations. *J. Prod. Agric.* **10**:314-319.

Ghosh, P. K. 2004. Growth, yield, competition and economics of groundnut/cereal fodder intercropping systems in the semi-arid tropics of India. *Field Crops Res.* **88**:227-237.

Hall, R. L. 1974a. An analysis of the nature of interface between plants of different species. I. Concepts and extension of the Dewit analysis to examine effects. *Aust. J. Agric. Res.* **25**: 739-747.

Hall, R. L. 1974b. Analysis of the nature of interface between plants of different species. II. Nutrient relations in a Nandi *Setaria* and green leaf *Desmodium* association with particular reference to potassium. *Aust. J. Agric. Res.***25**:749-756.

Harper, J. L. 1977. Population biology of plants. Orlando, Florida: Academic Press. 892 p.

Hiebsch, C. K. 1978. Interpretation of yields obtained in crop mixture. Abstracts of American Society of Agronomy, Madison, Wisconsin, pp.41.

Jain, T. C. and Rao, G. N. 1980. Note on a new approach to analysis of data in intercropping systems. *Ind. J. Agric. Sci.* **50**(12):970-972.

Khan, M. A. H., Sultana, N., Akhtar, S., Akter, N. and Zaman, M. S. 2017. Performance of intercropping groundnut with sesame. *Bangladesh Agron. J.* **20** (1): 99-105.

Maitra, S., Ghosh, D. C., Sounda, G. Jana, P.K. and Roy, D.K. 2000. Productivity, competition and economics of intercropping legumes in finger millet (*Eleusine coracana*) at different fertility levels. *Ind. J. Agric. Sci.* **70(**12**)**:824-828.

Maitra, S.,Samui, S. K., Roy, D. K. and Mondal, A. K. 2001**.** Effect of cotton based intercropping system under rainfed conditions in *Sundarban* region of West Bengal. *Ind. Agric.* **45 (**3-4**):** 157-162.

Maitra, S., Shankar, T. and Banerjee, P. 2020. Potential and advantages of maize-legume intercropping system (Online First), Intech Open, (*In:Maize - Production and Use, Ed.* Akbar Hossain), DOI: 10.5772/intechopen.91722. Available at: www.intechopen.com (Accessed on 26 March, 2020).

Manasa, P., Maitra, S. and Reddy, M. D. 2018. Effect of Summer Maize-Legume Intercropping System on Growth, Productivity and Competitive Ability of Crops. *Int. J. Eng. Res. Manag. Technol.* **8**(12): 2871-2875.

McGilchrist, C. A. 1965. Analysis of competition experiments. *Biometrics*. **21**:975–985.

Mead, R. and Willey, R.W. 1980. The concept of a "land equivalent ratio" and advantages in yields from intercropping. *Exp. Agric.* **16**:217-228.

Ofori, F. and Stern W.R. 1987. Cereal-legume intercropping systems. *Adv. Agron.* **40**:41-90.

Panda, S. K., Maitra, S., Panda, P., Shankar, T., Pal, A., Sairam, M. and Praharaj, S. 2021. Productivity and competitive ability of rabi maize and legumes intercropping system. *Crop Res.* **56** (3 & 4):98-104; DOI:10.31830/2454-1761.2021.016.

Sarkar, R. K. and Shit, D. 1993. Effect of inter cropping cereal, pulse, oilseed crops in redgram on yield, competition an advantage. *J. Agron. Crop Sci.* **170**: 171-176.

Sarkar, R. K., Shit, D. and Maitra, S. 2000. Competition function, productivity and economics of chickpea (Cicer arietinum)-based intercropping systems under rainfed conditions of Bihar plateau. *Ind. J. Agron.* **45**:681–688.

Schultz, B. B., Phillips, C., Rosset, P. and Vandermeer, J. 1982. An experiment in intercropping tomatoes and cucumbers in southern Michingan, USA. *Sci. Hortic*. **18**:1-8.

Spitters, C. J. T. and Van den Bergh, J. P. 1982. Competition between crops and weeds: a systems approach. Pages 137-148 in Biology and ecology of weeds (Holzner, W. and Numata, N., *Eds*.). The Hague, Dr. W Junk Publishers.

Willey, R. W. and Rao, M. R. 1980. A Competitive Ratio for quantifying competition between intercrops. *Exp. Agric*. **16**: 117-125.

Willey, R. W. 1979. Intercropping- its importance and research needs. Part 1: Competition and yield advantages. *Field Crop Abst*. **32**:1-10.

Willey, R. W. and Osiru, D. S .O. 1972. Studies on mixtures of maize and beans (*Phasrolus vulgaris*) with particular reference to plant population. *J. Agric. Sci. Cambridge*. **79**:519-529.

Willey, R. W., Natarajan, M., Reddy, M. S., Rao, M.R., Nambiar, P. T. C., Kannaiyan, J. and Bhatnagar, V. S. 1983. Intercropping studies with annual crops. *In*: Better crops for food, vol. 97. CIBA Foundation Symposium. pp. 83–100.

Yilmaz S, Atak M, Erayman M. 2008. Identification of advantages of maize-legume intercropping over solitary cropping through competition indices in the East Mediterranean Region. *Turk. J. Agric. For*. **32**(2):111-9.

3

Management of Intercropping System

Agriculture is the mainstay of the economy in different developing countries and it is not only nourishing the population, but also contributing a major share in country's gross domestic product (GDP). During the second half of the twentieth century and after, modern agriculture encouraged intensive cropping with high cost and consumption of fossil-fuel based energy. Yield enhancement was achieved by adoption of modern technologies, but that was at the cost of ecological unbalance. Presently, all are in search of agricultural sustainability. Sustainable agriculture ensures optimum and efficient use of resources with proper ecological balance targeting production for a long period with further improvement of resources and quality of life (Mousavi and Eskandari, 2011). Farming activities should be directed by following ecological principles and assuring higher ecosystem services targeting long term benefits. Sustainable agriculture ensures diversity in crop ecology and mono-cropping limits it, whereas intercropping creates above and below ground diversity with efficient use of resources.

Intercropping is an age-old practice and within the intercropping system, diversity is there in terms of crop choice, variety, planting geometry and spacing, proportion of crops and so on. Under different conditions, farmers adopt various intercropping practices and thus, an intercropping system itself becomes complicated. Two or more crops are grown together in intercropping and to obtain the considerable yield output efficient management and favourable environment are to be provided to all component crops as they can express their best. In this chapter, management of the intercropping system has been discussed.

3.1 Seedbed Preparation

Seedbed preparation is the basic step for successful growing of a crop. Firstly, initial tillage is performed under good tilth condition to make the soil favourable for seed germination and seedling vigour. The soil is to be prepared in such a manner that seeds sown come to contact of soil and the moisture available in soil helps the seed to germinate and later soil provides essential nutrients to plants. For small seeds, the soil should be tilled properly and made pulverized as it can facilitate necessary support to seeds. In intercropping, two or more crops are sown and mostly these crops are chosen from physiologically and morphologically different plant species to enhance resource utilization. Therefore, there will be difference in bed preparation prior to seeding in intercropping based on the crops planned. If two shallow rooted or two deep rooted crops are selected for intercropping, there will be no problem. But a combination of shallow and deep-rooted crops in the intercropping system, makes the situation critical. Considering the requirement of deep-rooted crop, the deep-tillage should be done initially followed by pulverization and bed preparation are to be done before seeding. There are some crops which are sown on ridges such as maize, sorghum, cotton etc. and priority is to be given for ridge making if these crops are selected in intercropping. Similarly, crops like wheat, mustard, pulses etc. are seeded on flat beds and combination of these crops in intercropping can easily be sown on flat beds. But when there will be crops with dissimilar preference, priority is to be given to base crop or economically important crop species. In major cases, both the crops are sown on flat beds and if possible, earthing-up is done to ridge preferring crop based on planting geometry adopted. But when sugarcane is considered as a main crop in polyculture, setts are planted in furrows, however, intercrop component is seeded on ridges.

3.2 Choice of Crops and Varieties

The advantages of polyculture have since long been realized (Darwin, 1859), but its spread is confined to subsistence farming in the developing countries (Altieri, 2004). Intercropping is recognized to increase yield output by 22–30% (Martin-Guay *et al.*, 2018) with a higher productivity stability (Raseduzzaman and Jensen, 2017). Actually, intercropping is a complex cropping system and it needs proper attention from the very beginning. The choice of crops and their varieties are of great concern because the success or failure greatly depends on it. The crops selected in mixed stands must have the inter-species complementary effect with a minimum competition among themselves for the limiting resources. Further, difference in maturity period of component crops should be about a month as after harvest of the early matured crop, the late maturing crop can enjoy the available resources fully. For example, the duration of maize hybrids is more than 110 days and if green gram is intercropped in maize, the pulse is harvested within 80 days or earlier. Moreover, the combination of short and long duration species, and deep and shallow-rooted crops in polyculture perform better.

The crop improvement programme carried out for several years has resulted in development of improved varieties and hybrids. But, a majority of cultivars developed are based on other needs. But, considering the need of intercropping in agricultural sustainability, breeders should look into the development of suitable varieties also (Lithourgidis *et al.*, 2011). Initially, Francis (1981) suggested the need for development of suitable cultivars for multiple and mixed cropping. The requirement for development of suitable varieties has been highlighted earlier by Davis and Woolley (1993), but very limited initiatives have been taken in this direction. The crop varieties suitable for intercropping should be photo insensitive as they can be grown in different seasons (Francis, 1989). The crop varieties should be chosen for intercrops with thinner leaves, erect and less branching type that can tolerate shade to some extent. Considering some special requirements in mixed stands, development of ideotype for monocropping and intercropping should be aimed separately because both types are grown for different objectives. So far, most of the crop varieties considered for intercropping are not bred for the purpose, these are developed for monocropping (Brooker *et al.*, 2014). In this context, when a highly productive cultivar suitable for monocropping is considered in intercropping that may not perform well under the stress of inter-species competition. The complex interaction and utilization of limited resource by the companion crops are to be understood for development of the suitable ideotypes for the mixed stands (Trinder *et al.*, 2012). Actually, it is not an easy task and main focus of crop breeding is targeted the development of ideotypes for monoculture. In an article, Saxena *et al.* (2018) suggested some ways to develop suitable varieties for intercropping. Some simulation and crop model-based studies highlighted the desired traits for breeding of cultivars suitable for intercropping (Berghujis *et al.*, 2020). In the recent years, focus has been given on genomic selection for development of suitable crop ideotype for intercropping, because of outdated and complex procedure of phenotypic selection (Annicchiarico *et al.*, 2021).

3.3 Sowing and Plant Stand

Crop sowing and ensuring optimum plant stand are essential tasks and in intercropping, more than one crop are sown; therefore, the complexity of seeding makes the tasks difficult. The optimum plant stand is an important factor for desired productivity of crops. Generally, when intercropping is adopted, plant population and planting geometry are altered and seeding is done accordingly. In widely spaced crops (cotton, pigeon pea, maize etc.) under uniform row planting method, intercrops are sown in between two rows of base crop and additive series of intercropping is followed in such case. Further, more space can be created for intercrops by adoption of paired row sowing of base crop and in such case, comparatively more plants for intercrops can be accommodated. The base crop population remains the same as it is maintained in the uniform row planting. Actually, in the paired row planting method, availability of more area for intercrops facilitates less competition between the crop species grown. Some experimental results revealed that paired row seeding yielded more than uniform row. Maitra *et al.* (2001) recorded that paired row cotton + green gram/ groundnut (2:1) produced more seed cotton yield and expressed yield advantage in coastal Sundarbans, West Bengal. Kakon *et al.* (2007) observed more yield of maize and higher benefit-cost ratio when paired row maize and green pea were intercropped (2:4) in Bangladesh. Similarly, Kumar *et al.* (2018) noted that in paired row seeding of pearl millet, intercropping of red gram and green gram recorded benefits of intercropping in terms of productivity of crops compared to uniform row seeding in the dry zone of Karnataka. However, the planting geometry of a replacement series is totally different where a crop is seeded for one or some rows and it is replaced by one or few rows of component crop(s). Sometimes, seeding rate of component crops are enhanced (even by adoption of closer row spacing) in the replacement series to obtain greater total yield.

Actually, the row proportion to be maintained greatly depends on the farmers' requirement and expected yield of the crop species chosen. Other agronomic practices such as tillage and bed preparation, seed treatment, gap filling etc. are adopted as per the recommended package of practices. The plant population maintained in an intercropping system affects the productivity of crops. Rao and Willey (1983) recorded greater LER with 70 thousand plants per ha in sorghum + red gram + sorghum (1:1:1, 45cm apart) sown in broad beds (150 cm); but when sorghum + red gram + sorghum (1:2:1, 30cm apart) or red gram + sorghum + red gram (1:2:1, 30cm apart) were grown, more LER was noted with 40 thousand plant population per hectare.

3.4 Fertilizer Application

Plant nutrient management is an essential operation for crop production, but the same in intercropping is more complex because of mixed crop species than monoculture. In general, intercropping removes more nutrients as more biomass is produced and hence, sufficient nutrients are to be provided. But, cereal–legume combination, a very common intercropping practice, requires less nitrogen because a portion of atmospheric N fixed by legumes in symbiotic association with Rhizobium is shared with cereals. In a study, Adu-Gyamfi *et al.* (1996) observed that application of 50 kg N per ha produced more grain yield of sorghum and red gram than 100 kg N per ha applied to sole sorghum and the former treatment also recorded higher plant nitrogen accumulation. A recent study indicated that reduced N fertilization increased N translocation and nitrogen harvest index in maize + pea intercropping system in northwestern China (Hu *et al.*, 2018). In another study, Kugbe *et al.* (2019) recorded that intercropping combination of maize + soybean with 50% recommended dose of fertilizer (RDF) resulted in superior benefit cost ratio to the treatment with 100% RDF in the guinea Savanna agro-ecological system of northern Ghana.

Cereals demand more N and in cereal-legume association, a major portion of applied N is absorbed by cereal; however, most of the legumes need only a small quantity of N at the early growth stages and use P for nodulation. After the nodulation takes place, the root exudates of legumes and beneficial microbes of rhizosphere make P available in the soil. Thus, both the crop components are benefited. In the soils with deficit nitrogen, legumes fix more nitrogen than the soils with more available nitrogen. In cereal + legume intercropping system, nitrogen is top-dressed for cereals and it is applied close to cereal rows. In an intercropping combination of more and less nutrient demanding crops, a nutrient dose of base crop is applied in a mixed stand. In some cases, fertilizer dose is adjusted based on the proportionate area occupied by each species (Sarkar *et al.*, 2000).

In the soils with micronutrient deficiency, especially molybdenum, basal application of micronutrients is preferred because molybdenum helps in nodulation of legumes (Alam, 2019). Also, Xue *et al.* (2016) reported that legumes-based intercropping enhances zinc and iron availability in soil. Actually, the belowground biological processes influence soil chemical and physical properties when legumes are considered as intercrops and these phenomena pronounce the complementary interaction among the crops in mixed stands (Maitra *et al.*, 2021).

3.5 Water Management

Water is an essential and precious input as well as natural resource for agricultural production. In spite of other uses of water, worldwide agriculture alone utilizes about 70% of available fresh water (Maitra and Pine, 2020). In the present consequences of global warming and climate change, weather aberration is causing water shortage in many regions hampering agriculture. Considering the above issues, focus has been shifted to low-input agriculture for maximization of water productivity under different conditions. The impact of non-judicious use of water may be fatal and hence, it is the need of the hour for planning of cropping system requiring low water. In this regard, intercropping has enough potential to fulfill the needs. But water management in intercropping is not as dissimilar crops are chosen and they coexist for a considerable period of their cycles and therefore, special care is to be taken to manage water requirement of crops. In additive intercropping, base crop is seeded with its desired population as adopted in pure stand and additionally intercrops are seeded. The combined plant-stand increases transpiration losses of water from the soil, but evaporation is reduced because of maximum ground coverage. Further, a soothing condition prevails at canopy level which may be beneficial for crop growth. Thus, water requirement for intercropping is more than sole cropping, but the difference is very marginal (Morris and Garrity, 1993); however, the advantage of intercropping is growing more crops in a unit area with greater biomass production. An experiment conducted in north-west China on water management in wheat + maize intercropping revealed that alternate irrigation exhibited higher water use efficiency (WUE) than conventional irrigation of either of the crops (Yang *et al.*, 2010). The strip-intercropping of pea in maize resulted in complementarity in sharing water and more productivity of crops. In a study, Nyawade *et al.* (2019) recorded a reduction of soil temperature (7.3°C at 0-30 cm depth) indicating more soil water content and crop water productivity than pure stand of potato in Kenya. In Egypt, El-Sherif and Ali (2015) recorded maize + soybean intercropping exhibited the more LER, combined yield and highest water use efficiency under irrigation with 100% of ET_0. Water consumption can be minimized by reduced tillage and mulching in wheat-maize intercropping (Yin *et al.*, 2015). Postponement of N topdressing reduced evaporation loss of soil moisture by 15–30% than the recommendation of topdressing (Teng *et al.*, 2016).

As it has been mentioned that dissimilar crops grown in intercropping may vary the water needs and full-fledged irrigation to more water-demanding crop may harm less water requiring crop. In this regard, maize + green gram may be a suitable example. Under conventional system, it is better to schedule irrigation based on IW/CPE (Irrigation Water / Cumulative Pan Evaporation) ratio. Moreover, there is scope for adoption of micro-irrigation with controlled water discharge to the crop species as per their requirements. The technological developments brought an arena of smart irrigation where sensor and IOT (Internet of Things)-based irrigation can be provided to crops in row and strip intercropping as per the crop needs. In the context of efficient soil moisture utilization, the concept of bio-irrigation may also be considered where the deep-rooted crop species moistens the rhizosphere and shallow-rooted component crop avails of the same (Saharan *et al.*, 2018; Singh *et al.*, 2020).

3.6 Weed Management

Weeds are harmful to crops as they compete with crops for nutrients, water and natural resources, suppress crop growth and productivity, and play a harmful role as hosts of insect-pests and pathogens. Weed management is a vital agronomic measure to obtain target yield and it is applicable to all forms of cropping systems, namely, monocropping, sequential cropping and intercropping. It is known that integrated weed management is the ideal practice to maintain the weeds population below the threshold level. Herbicides application is a cost-effective measure for weed management and in monocropping and sequential cropping the chemical weed management is widely practiced. But chemical weed control is not an easy task for an intercropping system where two or more crops are grown together. For example, when cereal + legume (monocotyledonous and dicotyledonous plants, respectively) combination is considered in an intercropping system, it is very difficult (not impossible) to choose suitable herbicide, though herbicides have selectivity. In general, greater area of land is covered by crop species grown in mixed stands than monocropping, and hence, weeds are suppressed (Liebman and Dyck, 1993; Bilalis *et al.*, 2010; Kithan and Longkumer, 2016; Divya *et al.*, 2020). Some legumes such as black gram, cowpea, green gram when grown in intercropping, automatically weeds population and weed growth are reduced because the legumes cover the land quickly suppressing the weeds growth.

In intercropping, the weed suppression ability greatly depends on crop choice, varieties used, seeding ratio, planting geometry and spatial arrangement of crops, plant stands, soil fertility, nutrients (more especially N fertilizer) application and soil moisture status. Manual weeding is a common practice in intercropping and at the early growth stage of the crops the field is hand weeded. Generally, there is no need for further weeding in intercropping when the canopy of crops is fully developed. In a study, Naher *et al.* (2018) observed that two hand weeding at 20 and 40 days after emergence reduced weeds density and weeds dry weight in maize + pea intercropping system and the treatment also resulted in higher LER and productivity of crops.

Based on the literature available, a list of herbicides that can be applied in the intercropping system has been presented below (Table 5.1).

Table 3.1. Herbicides used in intercropping system

Intercropping system	Herbicides (dose in kg ai ha^{-1})	Reference
Wheat + chickpea	Isoproturon (1.0)	Reddy, 2011
Wheat + mustard	Isoproturon (1.0)	Reddy, 2011
Maize + mung	Butachlor (1.25)	Reddy, 2011
Maize + cowpea	Alachlor (1.5), Alachlor (2.0), Metolachlor (1.0)	Reddy, 2011; Chalka and Nepalia, 2006
Maize + soybean	Pendimethalin (1.5), Benthiocarb (1.5); Alachlor (2.0), Metolachlor(1.0)	Verma and Dutta, 1984; Chalka and Nepalia, 2006
Sorghum + legumes	Linuron (0.25), Nitrofen (0.5) and Fluchloralin (0.25)	Abraham and Singh, 1984
Sorghum + blackgram	Alachlor (1.5)	Reddy, 2011
Sorghum + lablab	Dinitramine (0.75)	Reddy, 2011
Sorghum + pigeon pea	Ametryn (1.5) / Prometryn (1.0)/ Terbutryn (1.0)	Reddy, 2011
Pigeonpea + groundnut	Pendimethalin (1.0) / Fluchloralin (1.0)	Reddy, 2011

Integrated management of weeds can also be adopted where pre-emergence application of herbicides followed by mechanical management results in reduction of weeds density and weed biomass. In maize + cowpea and maize + black gram intercropping system, with application of pendimethalin at 0.75 kg/ha followed by one rotary hoeing on 35 DAS weeds growth was controlled (Rajeshkumar *et al.*, 2017; Rajeshkumar *et al.*, 2018). In another study based on maize and black gram intercropping system, pendimethalin 0.75 kg per ha as pre-emergence application + one manual weeding at 25 days after sowing checked weeds growth (Rahimi *et al.*, 2019).

3.7 Pest and Disease Management

The low insect-pest attack in intercropping is well recognized since long back (Nickel, 1973). Altieri *et al.* (1978) recorded reduced populace of *Empoasca krameri* by 26% in beans and *Spodoptera* by 14% in maize in intercropping than sole cropping of both the crops. In mixed stands, presence of beneficial insects are observed more checking the harmful pests (Maitra and Ray, 2019). Earlier researches evidenced that in maize + beans/ cowpea intercropping system, natural enemies were found more (Kyamanywa and Tukahirwa, 1988). Gavarra and Raros (1975) recorded greater numbers of predatory spiders when groundnut intercropped with maize than pure stands of maize. Reduced damage by weboorm (*Antigastra* sp.), a pernicious pest of sesame, was recorded when it intercropped sorghum in Nigeria (Litsinger and Moody, 1976). Adoption of cowpea + cotton intercropping is known to reduce sucking pests (Chikte *et al.*, 2008). The population of green stink bug (*Nezara viridula*) and stem borer (*Chilo zacconius*) were reduced in the upland rice + groundnut intercropping system (Epidi *et al.*, 2008). Earlier the research indicated that intercropping saves the target crop because of several mechanisms. The organic chemicals (may be volatile) emitted by the non-host crops grown in intercropping or offensive smells of natural enemies adversely affect the pests and provide some degree of protection (Song *et al.*, 2013; Dassou and Tixier, 2016; Sulvai *et al.*, 2016). Sometimes in mixed stand, some crop acts as a barrier crop which hinders the movements of insect pests and thus the susceptible plants suffer less. Success for intercropping for pest management depends on the choice of associated crops and their additional valuation after harvest, to some extent knowledge of the farmers and mechanization practice used (Lopes *et al.*, 2016). Like monoculture, integrated pest management practices are to be adopted in intercropping also. As the functional diversity in the intercropping system is created (Finckh *et al.*, 2000), pest population remains less than monoculture; hence, comparatively less efforts are needed for pest management.

In case of occurrence of diseases, the intercropping system was found to be better with less infection of harmful organisms. In wheat + *faba* bean intercropping system, there was reduced powdery mildew disease of wheat (Hauggaard-Nielsen *et al*., 2008). The incidence of wheat leaf rust (*Puccinia triticina*) and stripe rust (*Puccinia striiformis*) were also declined in wheat + rye intercropping system in China (Chen *et al*., 2007). In Denmark, the brown spot disease of lupin (*Pleiochaeta setosa*) was reduced in intercropping lupin + barley (Boudreau, 2013). Sorghum when intercropped with groundnut reduced groundnut bud necrosis disease (Narayanaswamy *et al*., 1988; Maitra *et al*., 2021). Moreover, incidence of various diseases was reduced by adoption of different intercropping combinations as recorded earlier such as bacterial wilt (*Pseudomonas solanacearum*) of potato in maize + potato intercropping, chocolate spot of *faba* bean (*Botrytis fabae*) in maize/ barley + *faba* bean combination, and Ascochyta blight of pea (*Mycosphaerella pinodes*) in cereals + pea intercropping system (Maitra *et al*., 2021). For management of diseases in intercropping system, integrated disease management approaches are to be adopted inclusive of choice of crop disease tolerant or resistant cultivars, adoption of appropriate rotation, seed treatment, removal of affected plant parts, clean cultivation, need based fungicides application etc.

Undoubtedly, management of intercropping system is little difficult than monocropping. But as intercropping system exhibits the multiple benefits as well as agricultural sustainability, it can be be considered as a suitable option for food and livelihood security of small farmers.

References

Abraham, C. T. and Singh, S. P. 1984. Weed management in sorghum-legume intercropping systems. *J. Agric. Sci. Camb.* **103**: 103-115.

Adu-Gyamfi, J. J., Katayama, K., Gayatri Devi, Rao, T. P. and Ito, O. 1996. Improvement of Soil and Fertilizer Nitrogen Use Efficiency in Sorghum/Pigeonpea Intercropping. In book: Ito, O.; Johansen. C., AduGyamfi, J. J., Katayama, K, Kumar Rao, V. D. K., Rego, T. J. (Eds.), Dynamics of roots and nitrogen in cropping systems of the Semi-Arid Tropics. Japan International Research Center for Agricultural Sciences, pp.493-506.

Alam, F., Kim, T. Y., Kim, S., Alam, S., Pramanik, P., Kim, P. J. and Lee, Y. B. 2019. Effect of molybdenum on nodulation, plant yield and nitrogen uptake in hairy vetch (*Vicia villosa* Roth). *Soil Sci. Plant Nutri.* **61**(4):1-12. DOI: 10.1080/00380768.2015.1030690.

Altieri, M. A. (2004). Linking ecologists and traditional farmers in the search for sustainable agriculture. *Front. Ecol. Environ.* **2**:35–42. doi: 10.1890/1540-9295 (2004)002 [0035: LEATFI]2.0.CO;2.

Altieri, M. A., Francis, C. A., Schoonhoven, A. V. and Doll, J. D. 1978. A review of insect prevalence in maize and bean polycultural systems. *Field Crops Res.* **1**:33-49.

Annicchiarico, P., Nazzicari. N., Notario, T., Monterrubio Martin, C., Romani, M., Ferrari, B. and Pecetti, L. 2021. Pea breeding for intercropping with cereals: variation for competitive ability and associated traits, and assessment of phenotypic and genomic selection strategies. *Front. Plant Sci.* **12**:731949. doi: 10.3389/fpls.2021.731949.

Berghujis, H. N. C., Wang, Z., Stomph, T.J., Weih, M., Van der Werf, W. and Vico, G., 2020. Identification of species traits enhancing yield in wheat-faba bean intercropping: development and sensitivity analysis of a minimalist mixture model. *Plant and Soil*, **455**(1): 203-226.

Bilalis, D., Papastylianou, P., Konstantas, A., Patsiali, S., Karkanis, A. and Efthimiadou, A. 2010. Weed-suppressive effects of maize-legume intercropping in organic farming, *Int. J. Pest Manag.* **56**:173-181.

Boudreau, M. A. 2013. Diseases in intercropping systems. *Annu. Rev. Phytopathol.* **51**:499–519.

Brooker, R. W., Bennett, A. E., Cong, W F., Daniell, T. J., George, T. S., Hallett, P. D., Hawes, C., Iannetta, P. P. M., Jones, H. G., Karley, A. J., Li, L., McKenzie, B. M., Pakeman, R. J., Paterson, E., Schob, C., Shen, J., Squire, G., Watson, C. A., Zhang, C., Zhang, F., Zhang, J. and White, P. J. 2014. Improving intercropping: a synthesis of research in agronomy, plant physiology and ecology. *New Phytol.* **206**(1):107-17. doi: 10.1111/nph.13132.

Chalka, M. K. and Nepalia, V. 2006. Nutrient uptake appraisal of maize intercropped with legumes and associated weeds under the influence of weed control. *Ind. J. Agric. Res.* **40**(2):86 – 91.

Chen. Y., Zhang, F., Tang, L., Zheng, Y. and Li, Y. 2007. Wheat powdery mildew and foliar N concentrations as influenced by N fertilization and belowground interactions with intercropped faba bean. *Plant Soil.* **291**:1–13.

Chikte, P., Thakare, S. M. and Bhalkare, S. K. 2008. Influence of various cotton-based intercropping systems on population dynamics of thrips, *Scircothrips dorsalis* Hood and whitefly, *Bemisia tabaci* Genn. *Res. Crop.* **9**:683-687.

Darwin, C. 1859. The origin of species. London: J. Murray.

Dassou, A. G. and Tixier, P. 2016. Response of pest control by generalist predators to local-scale plant diversity: a meta-analysis. *Ecol. Evol.* **6**:1143-1153. https://doi.org/10.1002/ece3.1917.

Davis, J. H. C. and Woolley, J. N. 1993. Genotypic requirement for intercropping, *Field Crops Res.* **34**:407-430.

Divya, R. K., Behera, B. and Jena, S. N. 2020. Effect of planting patterns and weed management practices on weed dynamics and nutrient mining in runner bean (*Phaseolus coccineus* L.) + maize (*Zea mays* L.) intercropping. *Int. J. Che. Stu.* **8**(1): 2704-2712.

El-Sherif, M. A. and Ali, M. M. 2015. Effect of deficit irrigation and soybean/maize intercropping on yield and water use efficiency Ahmed. *Int. J. Cur. Microbiol. and App. Sci.* **4**(12): 777-794.

Epidi, T. T., Bassey, A. E. and Zuofa, K. 2008. Influence of intercrops on pests' populations in upland rice (*Oryza sativa* L.). *Afr. J. Environ. Sci. Technol.* **2**:438-441.

Finckh, M. R., Gacek, E. S., Goyeau, H., Lannou, C., Merz, U., Mundt, C. C., Munk, L., Nadziak, J., Newton, A. C., de Vallavieille-Pope, C. and Wolfe, M. S. 2000. Cereal variety and species mixtures in practice, with emphasis on disease resistance. *Agronomie.* **20**:813-837.

Francis, C.A. 1989. Biological Efficiencies in Multiple-Cropping Systems. *Adv. Agron.* **42**:1- 42.

Francis, C.A., 1981. Development of plant genotypes for multiple cropping systems. *Plant Breeding.* **2**:179-232.

Gavarra, M. R. and Raros, R. S. 1975. Studies on the biology of predatory wolf spider, *Lycosa psedoannulata*. *Phil. Entomol.* **2**:427-444.

Hauggaard-Nielsen, H., Jørnsgaard, B., Kinane, J. and Jensen, E. S. 2008. Grain legume-cereal intercropping: the practical application of diversity, competition and facilitation in arable and organic cropping systems. *Renew. Agric. Food Syst.* **23**:3–12.

Hu, F., Tan, Y., Yu, A., Zhao, C., Coulter, J.A., Fan, Z., Yin, W., Fan, H. and Chai, Q. 2018. Low N fertilizer application and intercropping increases n concentration in pea (Pisum sativum L.) grains. *Front. Plant Sci.* **9**:1763. doi: 10.3389/fpls.2018.01763.

Kakon, S. S., Saha, R. R., Ahmed, F. and Jahan, M. A. H. S. 2007. Maize-pea intercropping as influenced by planting system and row arrangement. *J. Bangladesh Agril. Univ.* **5**(1): 37-41. ISSN 1810-3030.

Kithan, L. and Longkumer, T. L. 2016. Effect on yield and weed dynamics in maize *(Zea mays* L.) based intercropping systems under foothill condition of nagaland. *Int. J. Econ. Plants.* **3**(4):159-167.

Kugbe, J. X., Addai, I. K., Anyetin-Nya. and Asekabta, K. 2019. Intercropping and Fertilizer Rate Combinations Impact on Maize *(Zea Mays* L.) and Soybean *(Glycine Max* L (Merill)) Productivity: The Case Study in the Guinea Savannah Agro-Ecological Zone of Ghana. *J. Agri. Crops*. **5**(6): 87-99, DOI: https://doi.org/10.32861/jac.56.87.99.

Kumar, V., Singh, R. P., Kumar, S., Shukla, U. N. and Kumar, K. 2018. Performance of pearlmillet + greengram intercropping as influenced by different planting techniques and integrated nitrogen management under rainfed condition. *Int. J. Che. Stu.* **6**(3):705-708.

Kyamanywa, S. and Ampofo, J. K. O. 1988. Effect of cowpea/maize mixed cropping on the incident light at the cowpea canopy and flower thrips (Thysanoptera: *Thripidae*) population density. *Crop Prot.* **7**:186-189.

Liebman, M. and Dyck, E. 1993. Crop Rotation and Intercropping Strategies for Weed Management. *Ecol. Appl.* **3**(1):92-122.

Lithourgidis, A. S., Dordas, C. A., Damalas, C. A. and Vlachostergios, D. N. 2011. Annual intercrops: an alternative pathway for sustainable agriculture. *Aus. J. Crop Sci.* **5**(4):396-410.

Litsinger, J. A. and Moody, K. 1976. Integrated pest management in multiple cropping system. *In*: Ppaendick, R. I., Sanchez, P. A. and Triplett, G. B. (*Eds.*), Special Publication Number: 27, American Society of Agronomy, Madison, Wisconsin, USA, pp. 293-316.

Lopes, T., Hatt, S., Xu, Q., Chen, J., Liu, Y. and Francis, F. 2016. Wheat (Triticum aestivum L.) based intercropping systems for biological pest control. *Pest Manage. Sci.* **72**:2193-2202. https://doi.org/10.1002/ps.4332.

Maitra, S. and Pine, S. 2020. Smart Irrigation for Food Security and Agricultural Sustainability. *Ind. J. Nat. Sci.* **10**(60): 20435-20439.

Maitra, S. and Ray, D. P. 2019. Enrichment of biodiversity, influence in microbial population dynamics of soil and nutrient utilization in cereal-legume intercropping systems: A Review. *Int. J. Biores. Sci.* **6**(1):11-19. DOI: 10.30954/2347-9655.01.2019.3.

Maitra, S., Hossain, A., Brestic, M., Skalicky, M., Ondrisik, P., Gitari, H., Brahmachari, K., Shankar, T., Bhadra, P., Palai, J.B., Jena, J., Bhattacharya, U., Duvvada, S.K., Lalichetti, S. and Sairam, M. 2021. Intercropping- a low input agricultural strategy for food and environmental security. *Agronomy*. **11**(342):1–28.

Maitra, S., Samui, R. C., Roy, D. K. and Mondal, A. K. 2001. Effect of cotton based intercropping system under rainfed conditions in *Sundarban* region of West Bengal. *Ind Agric.* **45** (3-4): 157-162.

Martin-Guay, M. O., Paquette, A., Dupras, J., and Rivest, D. 2018. The new Green Revolution: Sustainable intensification of agriculture by intercropping. *Sci. Total Environ.* **615**: 767–772. doi: 10.1016/j.scitotenv.2017.10.024.

Morris, R. A. and Garrity, D. P. 1993. Resource capture and utilization in intercropping: water, *Field Crops Res.* **34**(3–4):303-317, https://doi.org/10.1016/0378-4290 (93)90119-8.

Mousavi, S. R. and Eskandari, H. 2011. A General Overview on Intercropping and Its Advantages in Sustainable Agriculture. *J. Appl. Environ. Biol. Sci.* **1**(11):482-486.

Naher, Q., Karim, S. M. R. and Begum, M. 2018. Performance of legumes on weed suppression with hybrid maize intercropping. *Bangladesh Agron. J.* **21**(2): 33-44.

Narayanaswamy, P., Ganghadharan, K., Chandrasekharan, G., Velazhagan, R. and Karunanidhi, K. 1988. In Proceedings of National Workshop on Pests and Diseases, Tamilnadu Agricultural University, Tamilnadu, India, 16–18 September 1988.

Nickel, J. L. 1973. Pest situation in changing agricultural system: a review. *Bull. Entomol. Soc. Amer.* **54**:76-86.

Nyawade, S.O., Karanja, N. N., Gachene, C. K. K., Gitari, H. I., Schulte-Geldermann, E. and Parker, M. L. 2019. Intercropping Optimizes Soil Temperature and Increases Crop Water Productivity and Radiation Use Efficiency of Rainfed Potato. *Ame. J. Potato Res.* **96**:457–471, DOI: 10.1007/s12230-019-09737-4.

Rahimi I., Amanullah, M.M., Ananthi, T. and Mariappan, G. 2019. Influence of intercropping and weed management practices on weed parameters and yield of maize. *Int. J. Curr. Microbiol. App. Sci.* **8**(04): 2167-2172. doi: https://doi.org/10.20546/ijcmas.2019.804.254.

Rajeshkumar, A., Venkataraman, N. S. and Ramadass, S. 2018. Integrated weed management in maize-based intercropping systems. *Ind. J. Weed Sci.* **50**(1):79–81.

Rajeshkumar, A., Venkataraman, N. S., Ramadass, S., Ajaykumar, R and Thirumeninathan, S. 2017. A Study on inter-cropping system and weed management practices on weed interference and productivity of maize. *Int. J. Che. Stu.* **5**(5):847-851.

Rao, M.R. and Willey, R.W. 1983. Effects of pigeonpea plant population and row arrangement in sorghumlpigei-rnpea intercropping. *Field Crops Res.* **7**: 203-21 2.

Raseduzzaman, M. and Jensen, E. S. 2017. Does intercropping enhance yield stability in arable crop production? a meta-analysis. *Eur. J. Agron.* **91**:25–33. doi: 10.1016/j.eja.2017.09.009.

Reddy, S. R. 2011. Principles of agronomy. Kalyani Publishers, India. pp. 581.

Saharan, K., Schütz, L, Kahmen, A., Wiemken, A., Boller, T. and Mathimaran, N. 2018. Finger millet growth and nutrient uptake is improved in intercropping with pigeon pea through "biofertilization" and "bioirrigation" mediated by arbuscular mycorrhizal fungi and plant growth promoting rhizobacteria. *Fron. Environ. Sci.* **6**(46):1-11.

Sarkar, R. K., Shit, D. and Maitra, S. 2000. Competition functions, productivity and economics of chickpea (Cicer arietinum)-based intercropping system under rainfed conditions of Bihar plateau. *Ind. J. Agron.* **45** (4) : 681-686.

Saxena, K. B., Choudhary, A. K., Saxena, R. K. and Varshney, R. K. 2018. Breeding pigeonpea cultivars for intercropping: synthesis and strategies. *Breeding Sci.* **68**: 159–167, doi:10.1270/jsbbs.17105.

Singh, D., Mathimaran, N., Boller, T., Kahmen, A. 2020. Deep-rooted pigeon pea promotes the water relations and survival of shallow-rooted finger millet during drought—Despite strong competitive interactions at ambient water availability. *PLoS ONE.* **15**(2):e0228993. https://doi.org/10.1371/journal.pone.0228993.

Song, B., Tang, G., Sang, X., Zhang, J., Yao, Y. and Wiggins, N. 2013. Intercropping with aromatic plants hindered the occurrence of aphis citricola in an apple orchard system by shifting predator-prey abundances. *Biocon. Sci. Techno.* **3**: 381-395. https://doi.org/10.1080/09583157.2013.763904.

Sulvai, F., Chaúque, B. J. M. and Macuvele, D. L. P. 2016. Intercropping of lettuce and onion controls caterpillar thread, agrotis ípsilon major insect pest of lettuce. *Che. Biol. Technol. Agric.* **3**:28. https://doi.org/10.1186/s40538-016-0079-z.

Teng, Y., Zhao, C., Chai, Q., Hu, F. and Feng, F., 2016. Effects of postponing nitrogen topdressing on water use characteristics of maize-pea intercropping system. *Acta Agron. Sin.* **42**(3):446-455.

Trinder, C., Brooker, R., Davidson, H. and Robinson, D. 2012. Dynamic trajectories of growth and nitrogen capture by competing plants. *New Phytol.* **193**: 948–958.

Verma, S. P. and Dutta, B. N. 1984. Weed management studies in maize and soybean intercropping systems. *Aus. Weeds.* **3**(4): 140-145.

Xue Y., Xia, H., Christie, P., Zhang, Z., Li, L. and Tang, C. 2016. Crop acquisition of phosphorus, iron and zinc from soil in cereal/legume intercropping systems: a critical review. *Ann. Bot. Lond.* **117**:363–77. doi:10.1093/aob/mcv182.

Yang, C. H., Chai, Q. and Huang, G. B. 2010. Root distribution and yield responses of wheat/maize intercropping to alternate irrigation in the arid areas of northwest China. *Plant Soil Environ.* **56**(6):253–262.

Yin, W., Yu, A., Chai, Q., Hu, F., Feng, F., Gan, Y. 2015. Wheat and maize relay-planting with straw covering increases water use efficiency up to 46 %. *Agron. Sustain. Dev.* **35:**815–825 (2015). https://doi.org/10.1007/s13593-015-0286-1.

4

Advantages of Intercropping

During recent time, world agriculture is in search of sustainability in farming. However, it is tedious job for the developing countries because of poor input management and degradation of resources, ever-growing human population and increasing demand for food and livestock feed, and shrinkage of land and other natural resources. Further, issues related to global warming and climate change are making it more difficult due to weather aberrations. Under the circumstances, it is necessary to adopt a cropping system approach for efficient resource utilization. The system consists of inter-related and interacting components of the farm and the proper system approach scientifically interacts with all resources of the farm. But, development of a suitable and efficient cropping system considering resource base and agroclimatic feasibility is a tremendous job targeting potential yield. The outcome of a cropping system is measured by yield output of crop(s) grown with optimum utilization of resources. The recent agronomic perceptions tell us to estimate system efficiency by not only quantifying yield output of crops(s) but also to consider spatial and temporal allocation of crops grown in the cropping system.

The cropping system can be broadly divided into two categories, namely, sequential and intercropping. Exploitation of the potential of sequential cropping was started in India during Green Revolution era and over-exploitation of green revolution technologies (GRTs) caused fatal impacts on crop yield and resources. On the other hand, the age-old practice of intercropping remained neglected and under-exploited. Various reports and literature clearly revealed the advantages of intercropping over sole cropping, still it was unkempt probably because of the complexity of the system itself and adopters' ignorance. Some of the researchers went one step ahead and expressed their views that intercropping system is suitable for only small farmers. Further, enough research was also not carried out for development of suitable tools and machines to reduce the complexity of crop management in an intercropping system. But, the practice of polyculture for several centuries evidenced its potential in achieving agricultural sustainability. As agricultural sustainability is a prime concern of the world, hence, the potential of the intercropping system can be re-evaluated in that direction.

Intercropping system is known to grow more than one crop in a piece of land during the same period (Figure 4.1) and therefore, available soil moisture and nutrients, carbon dioxide and solar radiation are better exploited from unit area resulting in harvest of more biomass. In this way, intercropping system enhances efficient resource use, manages biotic agent population dynamics, conserves soil by covering maximum ground area, checks loss of soil and nutrients from top soil through runoff, and ensures greater ecosystem services (Maitra *et al.*, 2019, 2020, 2021). Further, when legumes are considered as components in an intercropping system, benefits are obtained in terms of enhancement of helpful microbes population in the soil and well as enrichment of microbial diversity, increase in N in post-harvest soil, N sharing by nonlegume components and the system facilitates overall soil health improvement (Maitra and Ray, 2019; Maitra *et al.*, 2021). All these beneficial activities are not only synonymous to crop yields enhancement from a unit area but also considered as a step ahead to achieve agricultural sustainability (Maitra *et al.*, 2021).

Figure 4.1. (A) Intercropping sesame and groundnut **(B)** Maize and cluster bean

In the fragile ecological conditions and dryland regions, crop failure is a very common phenomenon under aberrant climatic situations and monocropping is more vulnerable under the mentioned conditions. However, in intercropping, two or more crops grown together may not be equally affected by climatic abnormalities and hardy crop chosen in combination may yield to a certain extent; thus, intercropping has enough potential to ascertain natural insurance. In most of the cases, physiologically and morphologically dissimilar crops are chosen in an intercropping system that may create some limitations as highlighted by some of the researchers. But it is an opportunity for smallholders. As the crops grown in association may be matured at different periods and mechanized harvesting is difficult (except forage intercropping), smallholders can utilize their family labourers avoiding unemployment during lean period. Thus, the benefit of employment generation in resource poor areas can be considered as an important advantage. Moreover, poor and small farmers practicing subsistence farming can harvest diversified (cereal + legume) and nutritious foods for their own consumption which leads to food and nutritional security. The functional diversity created in growing of different crop species also benefits the crop protection against various biotic stresses. In the chapter, the advantages of intercropping systems towards achieving agricultural sustainability (Figure 4.2) have been presented considering all aspects.

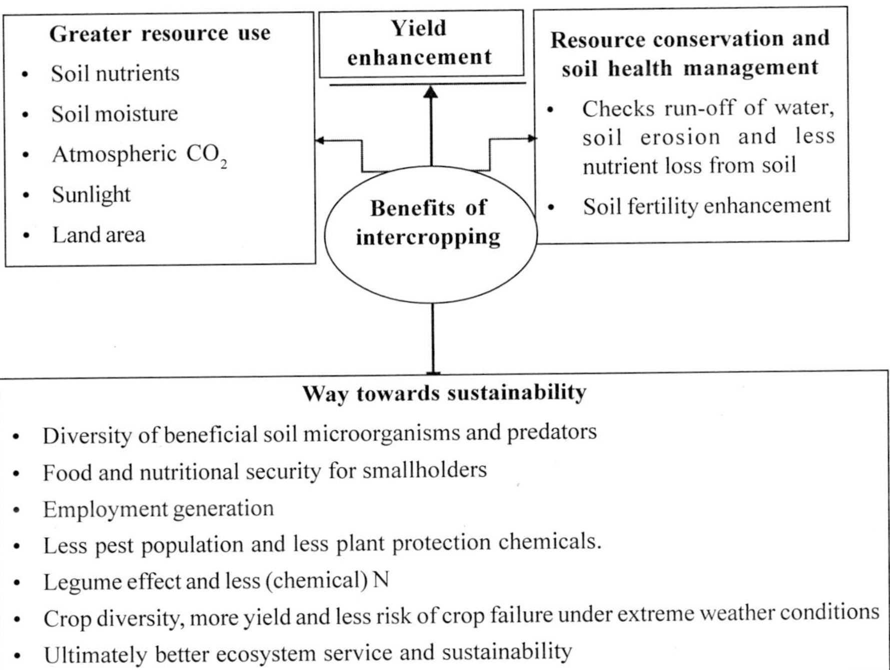

Figure 4.2. Advantages of intercropping system (Maitra *et al.*, 2021)

4.1 Yield Benefits

Two or more crops are cultivated in an intercropping system simultaneously (Gitari *et al.*, 2021) and many times the combined plant population exceeds the normal plant stand of pure crops. Choice of crop is a vital factor in the success of the intercropping system. There must be complementarity among the crop species grown in polyculture with a less competition to obtain yield advantages (Willey, 1979). In widely spaced crops, additive series of intercropping is generally preferred where addition of crop(s) takes place in normal stand of base crop and by getting normal population base crop yields very close to its pure stand; whereas additional yield comes from intercrop (Manasa *et al.*, 2018, 2019; Panda *et al.*, 2021). Further, when maximum complementarity expresses in intercropping, there is possibility of yield enhancement of base crop (Assefa *et al*., 2016). In a study in Ethiopia, Assefa *et al*. (2016) recorded that there was 10.9 percent yield enhancement when paired row maize + narrow leaf lupine was intercropped. If complementary effects are exhibited among plant species, in replacement series also more combined yield is achieved (Maitra *et al*., 2000, 2001). There are some competitive functions developed by earlier researchers of which monetary advantage and base crop equivalent yield, relative yield total (RYT) and relative value total (RVT) exhibit the yield advantages of intercropping system. A study carried out by Mandal *et al*. (2014) recorded greater maize equivalent yield (MEY) of 5.48 t/ha when soybean intercropped in maize (with a row proportion of 1 row maize + 2 rows of soybean) against pure stand of maize productivity of 2.48 t/ha in subtropical conditions in the red and lateritic belt of West Bengal, India. Similarly, Manasa *et al*. (2020) observed MEY of 7.6t/ha in maize + groundnut, however, the pure stand of maize yielded 5.7 t/ha of grains with a relative yield value of 1.47, indicating yield advantage.

4.2 Enhanced Resource Use

Accommodation of more than one crop in the same field simultaneously ensures greater use of available resources (Maitra *et al*., 2019), namely, land, soil water, available nutrients in soil, solar radiation and CO_2 etc. These resources are utilized by more plants or crops than pure stand. As the available resources are used by more plants than pure stand of individual crops in unit area, greater total biomass yield is obtained from the land. Higher biomass yield is directly proportional to maximum level of complementarity among the crops and the crops coexisting together should not compete for common resources simultaneously. It is also synonymous to weaker interspecies competition. In general, physiologically and morphologically dissimilar crops are grown in intercropping system and such complementarity (as well as less interspecies competition) is observed. Here, as example, maize + green gram can be considered. Maize demands more nutrients and resources after six weeks of sowing, while green gram completes more than 50 percent of its cycle by this period. On the other hand, maize is a more nitrogen demanding crop and green gram needs less nitrogen. So, these two crops exhibit a less competitive interaction; rather maize gets benefit from green gram in terms of sharing of biologically fixed nitrogen by the legume component. The combination of shallow-rooted and deep-rooted crops could be suitable example of intercropping where deep rooted crop absorbs nutrients and soil moisture from the deeper horizon of the soil and shallow-rooted crops penetrate up to 15-30cm depth of the soil. The other important factors in increasing resource use efficiency are plant population and planting geometry. Greater biomass yield from unit area by seeding more plants with suitable alteration in planting geometry can be obtained in intercropping system.

4.2.1 Improved Soil Properties

The cereal- legume mixture in intercropping is practiced worldwide in intercropping and the presence of legumes can contribute biologically fixed nitrogen ranging from 80-350 kg/ha based on legume species and duration (Peoples and Craswell, 1992). When legumes are incorporated in intercropping system, microbes population is changed and that increases beneficial microbes in rhizosphere facilitating enhanced nutrients mineralization (Mobasser *et al.*, 2014) along with reduction of pathogens species in soil (Maitra and Ray, 2019). Earlier researches evidenced that physico-chemical properties in soil are altered due to legumes incorporation in an intercropping system (Zhang *et al.*, 2004). Legumes as component crops influence soil physical properties (Srinivasarao *et al.*, 2012). The leaf fall and root biomass produced by legumes significantly contribute in qualitative enhancement of soil physical properties inclusive of soil organic matter (SOM) (Lal, 2015). Increase in SOM prevents soil from erosion and increases water holding capacity along with aggregate stability (Lithourgidis *et al.*, 2011; Mousavi *et al.*, 2009).

As the microbial population dynamics is changed by legumes-based intercropping systems, organic carbon content of soil is improved (Song *et al.*, 2007) and C/N ratio of soil is properly maintained (Dexter, 2004; Schädler *et al.*, 2011). The legumes can alter the essential nutrient availability to plants (Peyraud *et al.*, 2009) by releasing root exudates consisting of various enzymes and organic acids. Legumes play a valuable role on available nutrients, soil pH and soil organic carbon stock (Mugwe *et al.*, 2004). Further, legumes included in the intercropping system change the soil pH by releasing organic acids enhancing soil phosphorus availability (Buerkert *et al.*, 2000; Bado *et al.*, 2004). Further, legumes cause sequestration of atmospheric CO_2 (Turnbull and Bowman 2002; Pikul *et al.*, 2008; Lal, 2015). Legumes in the cropping system also play a magnificent role that they hold pesticides in soil and never let them carry forward to contaminate soil as well as water bodies preventing erosion and runoff to a greater extent (Fester *et al.*, 2014). Hydrogen, a byproduct of symbiotic nitrogen fixation, has bioactive property improving the abiotic stress tolerance of plants (Cui *et al.*, 2013). Other than biological N fixation, *Rhizobia* (associated with legumes) provide advantages in elimination of soil pollutants directly (Jin *et al.*, 2013), or indirectly enhance other degrading microbes and enzymes (Kaiya *et al.*, 2012), act in metal bioremediation (Fester *et al.*, 2014).

Improvement of soil biological properties is most important due to the presence of legumes in the intercropping system (Suman *et al.*, 2006). As N is a limiting nutrient of majority of the arable lands, legumes can undergo biological nitrogen fixation (BNF). Not only nitrogen but rhizobium facilitates availability of soil phosphorus to plants (Nuruzzaman *et al*., 2006; Palai *et al*., 2021). The addition of legumes in intercropping enhances the soil microbial activity as well as increases microbial diversity (Alvey *et al*., 2003). A small change in the concentration of nitrogen, cellulose, lignin content or C:N ratio of residue triggers the divergence of soil microbial status (Meena *et al*., 2014). Legume-rhizobium association also influence plant growth promoting microbes (Schelud'ko *et al*., 2009) and there is also report of tripartite symbiotic association of mycorrhiza-legume-rhizobium which further enhances the plant nutrition (Hayman 1986; Scheublin *et al*., 2004).

Earlier studies recorded that intercropping of pea and barley enhanced phosphorus, sulfur and potassium and increased crops productivity (Hauggaard-Nielsen *et al*., 2009). In intercropping, phosphorus uptake is increased by the crops (Ae *et al*., 1990; Cu *et al*., 2005; Mobasser *et al*., 2014). In problematic soil also, the intercropping system plays a significant role in soil nutrients dynamics. In acid soils, phosphorus fixation is a common phenomenon and root exudates released by legumes can cause availability of soil phosphorus (Li *et al*, 2013); therefore, intercropping legumes with other crop species is beneficial. Aluminum toxicity is a major constraint in acid soil and the root exudates and organic acids produced by legumes provide a safeguard to roots against aluminum toxicity (Ryan *et al*., 2011). In alkaline soil, intercropping of groundnut and maize recorded higher iron foraging (Dai *et al*., 2019). Xue *et al*. (2016) observed more availability of Fe and Zn in soil when cereal-legume was intercropped. Choice of crop species in intercropping is important to harness beneficial impacts in terms of improvement in soil health and enhanced soil nutrients dynamics.

4.2.2 Efficient Soil Moisture Use

In field conditions, plants absorb water and nutrients mainly from soil. Two or more different crops are cultivated in an intercropping system simultaneously. Plants grown together absorb the same available soil moisture from the field. In other words, it may be narrated that the same quantity of available soil moisture is used by more plants (as the combined plant population remains generally more in intercropping than pure stand). Further, a mixture of shallow and deep-rooted crops is advantageous for absorbing soil moisture from different layers. Such a combination enhances water use efficiency. Chen *et al*. (2018) recorded that strip-intercropping maize and pea registered greater water use efficiency with enhanced productivity. Moreover, Chai *et al*. (2014) observed an enhanced water use efficiency when maize and field pea intercropped than sole cropping of both the crops. In intercropping system, shallow-rooted plants are facilitated by bio-irrigation which is a hydraulic lift of soil moisture from deeper zone of the soil and its redistribution to different soil layers. Even under soil moisture stress conditions, the deep-rooted plants absorb soil moisture from deep and distribute some quantity to roots of drier parts in the upper layer. The water is further distributed to all parts of the root and the neighbouring shallow-rooted crop in association gets benefit. Experiments carried out on bio-irrigation revealed that pigeon pea performed as bio-irrigator when grown in polyculture with cereals (Saharan *et al.*, 2018; Singh *et al.*, 2020).

4.2.3 Efficient Use of Atmospheric CO_2

Plants produce assimilates in presence of sunlight and utilizes atmospheric CO_2 through the photosynthesis. More biomass production by plants in mixture is synonymous to utilization of more CO_2 and releasing more oxygen to atmosphere (Adler *et al.*, 2007; Signor and Cerri, 2013; Collins *et al.*, 2017). Thus, intercropping is considered as an environmentally friendly cropping system.

4.2.4 Solar Radiation Use

Crops need solar radiation for assimilate production as well as growth and productivity. In intercropping, there is advantage of inclusion of morphologically dissimilar crops. A combination of short and tall crops can harvest more solar energy than crops of equal height. Undoubtedly, in such a combination, the taller plant gets the advantage and the dwarf species becomes suppressed. But proper crop choice can solve the issue and the ideal combination should be the crops with different light requirements. Ghanbari *et al.* (2010) observed that maize + cowpea intercropping system registered more sunlight interception than pure stand of maize. Similarly, Mahallati *et al.* (2015) recorded that intercropping of maize + bean registered absorption of more solar radiation and yield than pure stand of both the crops. In a maize-based intercropping system, more photosynthetically active radiation (PAR) was used by the crops in mixed stand (Kermah *et al.*, 2017); but sole cropping of soybean and cowpea utilized more PAR than intercropping. Such results were obtained probably due to the taller canopy structure of maize which created partial shade to dwarf statured legumes in mixture. In an experiment in China, Raza *et al.* (2019) defoliated two leaves of maize from the top at silking stage and noted increased yields of intercropping soybean in maize because of more light interception and efficient dry matter partitioning.

4.2.5 Area Utilization

As two or more crops are grown simultaneously in a land in intercropping system, land area is better utilized. In additive series of intercropping system, 100% plant stand of base crop is maintained (as sown in sole cropping) with inclusion of some more crop(s). In replacement series of intercropping, an additional plant stand is maintained sometimes for all species other than sole crop (by increasing seed rate or closing spacing) to get advantages of intercropping. In this way, a unit area is occupied by more plants combinedly compared to sole cropping. But interspecies competition may be an issue in intercropping; hence, prime importance is to be given on selection of crops and they should be of dissimilar canopy stature, duration and period for maximum demand for resources to minimize the competition (Maitra *et al.*, 2019).

To evaluate the land use efficiency, the concept of Land Equivalent Ratio (LER) was developed by Willey and Osiru (1972) that denotes a total of proportions of productivity of each component crop to its respective pure stand yield. The LER does not consider the 'time' factor and it overestimates the efficiency of intercropping system which can be considered as a limitation of the expression. But another expression, named as area time equivalent ratio (ATER) includes the 'time' factor along with land area (Hiebsch, 1978). The LER or ATER values if appear as more than 1, clearly indicate yield advantages in intercropping systems (Table 4.1 and 4.2).

Table 4.1. LER of some legume and non-legume intercropping systems

Intercropping system	Proportion/ description	LER non-legume	LER legume	LER combined	References
Maize + soybean	Relay intercropping in strips	0.97	0.90	1.87	Raza *et al.*, 2020
Cotton + green gram	2:2	0.85	0.85	1.70	Khan *et al.*, 2020
Maize + groundnut	2:1	0.89	0.86	1.75	Panda *et al.*, 2021
Maize + groundnut	2:2	0.92	0.78	1.70	Panda *et al.*, 2021

Table 4.2. ATER of intercropping systems from research findings

Intercropping system	Proportion	ATER	Country	References
Sorghum + pearl millet	4:4	1.13	Pakistan	Amanullah *et al.*, 2020
Wheat + Faba bean	2:1	1.19	Pakistan	Amanullah *et al.*, 2020
Rice + Spinach	-	1.23	India	Habimana *et al.*, 2021
Maize + groundnut	2:3	1.49	India	Panda *et al.*, 2021
Roselle: cluster bean	1:3	1.16	Egypt	Mohammed *et al.*, 2022

4.3 Efficient Resource Conservation

An important benefit of intercropping is conservation of resources. As the maximum ground area is occupied by crops in mixed stand, it is possible to conserve soil and water. Further, nutrient loss from the top-soil is also checked due to adoption of intercropping.

4.3.1 Checking Run-off of Water, Soil Erosion and Nutrient Loss

Coverage of more ground area in intercropping system restricts free movement of excess water (i.e., runoff), and carrying out of soil and nutrients from the top-soil with runoff of water. Alley cropping in an agroforestry system can be an ideal example of checking soil and nutrients loss from the top-soil. Experiments carried out on alley cropping revealed that *Gliricidia* in alleys reduced run-off of water (by 28.2%) and loss of top-soil (by 49.3-51.1%) over the marginal land where alley cropping was not adopted (Madhu *et al.*, 2019). Moreover, the study recorded that alley cropping conserved soil nutrients, namely, organic carbon (63.4 kg/ ha), nitrogen (5.0 kg/ ha), phosphorus (0.3 kg/ ha) and potassium (2.4 kg/ ha). Another effective crop species for alley cropping is *Leucaena* that reduced runoff of water (by 18.3-18.7%) and soil erosion (by 37.2-43.0%) in the above study. Like *Gliricidia*-based alley cropping, *Leucaena* also conserved soil nutrients such as organic carbon (57.7 kg/ha), nitrogen (4.6 kg/ha), phosphorus (0.3 kg/ha) and potassium (2.2 kg/ha). Dass and Sudhishir (2010) recorded that intercropping black gram in finger millet checked runoff of water and loss of primary nutrients.

4.3.2 Soil Fertility Enhancement

Intercropping of cereals and legumes are known to increase soil fertility due to leaf fall and biological nitrogen fixation contributed by legumes (Garg, 2007; Maitra *et al.,* 2019). Research results recorded that intercropping of finger millet and pulses increased soil fertility (Dass and Sudhishir, 2010). The combination of cereal and legume in intercropping caused immobilization of nitrogen in soil (Iqbal *et al.*, 2018). Studies on nutrient balance resulted in increased nitrogen fertility of soil because of adoption of legume-based intercropping system. Intercropping maize + groundnut efficiently increases N content in post-harvest soil (Mucheru-Muna *et al.*, 2010; Choudhary and Choudhary, 2016). Intercropping soybean and groundnut in maize also increases phosphorus and potassium content in soil (Choudhary and Choudhary, 2016). The presence of legumes in crop combination is responsible for improvement of soil fertility in intercropping systems with a combination of cereals and legumes. Legumes perform BNF and the fixed N can directly be available to the current season non-legumes in mixture (Adeniyan *et al.*, 2007; Dahmardeh *et al.*, 2010); however, a portion of nitrogen is added to the soil which the succeeding crop can use as residual N availability (Raji 2007; Barbosae *et al.*, 2008).

4.4 Way Towards Sustainability

As the legumes have the potential to fix nitrogen biologically, incorporation of legumes in intercropping has also potential in reducing the use of nitrogenous fertilizers in the cropping system. Production of industrial nitrogenous fertilizer such as urea releases greenhouse gases (GHGs). An estimate revealed that to fulfill the urea requirement of the world, an amount of 300 Tg of CO_2 equivalent GHGs are emitted to the atmosphere (Jensen *et al.*, 2012). In the present context of global warming and climate change, excessive GHGs emission to the atmosphere is further creating another dimension in the vulnerability of sustainable agricultural production. Legumes are regarded for BNF and the unique phenomenon in intercropping of legumes with non-legume is that the fixed N is shared by legumes to non-legume. The feature of legume and non-legume combination ultimately keeps its presence not only in N-economy, but also its contribution in restricting pollution to the atmosphere. The soil with higher content in soil organic matter (SOM) is considered as a prime indicator of fertility that can maintain soil biological health. In an intercropping system with a combination of legumes and others, enriches SOM through leaf falls of legumes and biomass (as roots, stover and stubble incorporation); that ultimately improves SOM. In other words, we can say it is synonymous to SOC sequestration. The other advantages of intercropping are pronounced as diversification of above-and-below ground biotic agents such as crop species, weeds, insects and pathogens. The functional diversity is useful in various ways, namely for plant protection, creation of biodiversity and efficient ecosystem services. Further, the intercropping system has enough potential to ensure food, nutrition and livelihood security to small farmers practicing subsistence farming. Some important features related to the intercropping system and agricultural sustainability have been detailed below.

4.4.1 Enrichment of Biotic Diversity

Intercropping system offers biodiversity enrichment and it is visible in both the ways, above and below ground diversity of flora and fauna. Different crops grown simultaneously in the same field in an intercropping system ensures crop diversification. Besides, mixed stand of crops nurture wide range of birds, insects and arthropods. Afrin *et al.* (2017) recorded abundance of butterfly and natural enemies when mustard was intercropped with wheat and other crops. At below ground level, diversity of rhizospheric microbes were also recorded (Maitra and Ray, 2019). In any intercropping system, when legumes are chosen as crop species, automatically the Rhizobium population is enhanced. Not only the abundance of Rhizobium but also increased population of Betaproteo-bacteria, Alphaproteo-bacteria, Cyano-bacteria, *Pseudomonas* sp. were reported in the earlier studies (Spehn *et al.* 2000; Li and Wu, 2018). Increase in biodiversity is synonymous to a healthy ecosystem. Interestingly, harmful microbes population was decreased due to adoption of intercropping (Finckh *et al.*, 2000); thus, it favoured crop growth and productivity.

4.4.2 Ensures Food and Nutritional Security

In the developing world, most of the small farmers prefer subsistence farming for their food needs and livelihood. The nutritional requirement of the farm family is not properly taken care of because of limitid resources and capital. Thus, food and nutritional security are neglected leading to malnutrition which is great challenge. In that context, diversification of crops in the form of intercropping system with a mixture of pulses (legumes) and oilseeds with cereals can ensure a healthy and nutritious dietary intake for the target section of the farming community across the globe. The role of intercropping may be re-evaluated in this direction for achieving the target of SDG 2 (zero hunger).

4.4.3 Maintained Pest Population Dynamics

Intercropping system creates a functional diversity that leads to reduction of weeds, harmful insects and diseases. Two or more crops are grown in mixed stands with or without a distinct proportion creating the functional diversity in intercropping system. Further, the mixed stand of crops attracts the predators, natural enemies and pollinator insects that play beneficial roles in maintaining the pest population dynamics as well as enhancing the crop yield (Nicholls and Alteri, 2013; Maitra *et al.*, 2019). Earlier researches evidenced that intercropping *Brassicas* with other crops that are taxonomically different, reduced pest population and increased yield (Dempster and Coaker, 1974; Burn *et al.,* 1987). In maize-based intercropping system with cowpea, due to abundance of natural enemies pests population was comparatively low than pure stand of maize (Kyamanywa and Tukahirwa, 1988). Intercropping cowpea in cotton reduced sucking pest menace in cotton (Chikte *et al.*, 2008). An intercropping combination of upland rice and peanut minimized presence of green stink bug (*Nezara viridula*), a polyphagous pest, and stem borer (*Chilo zacconius*) population (Epidi *et al.*, 2008).

Not only insect pests, but also plant diseases are restricted because of the creation of diversity in mixed stands (Finckh *et al.*, 2000). An intercropping combination of sorghum and peanut reduced the bud necrosis in peanut (Narayanaswamy *et al.*, 1988). The following table (Table 4.3) showed positive influence of intercropping systems by reducing the disease incidence in crops.

Table 4.3. Disease reduction in intercropping system

Crop	Name of the restricted disease	Intercropping combination	Reference
Potato	Bbacterial wilt (*Pseudomonas solanacearum*)	Maize + potato	Autrique and Potts,1987
Faba bean	Chocolate spot (*Botrytis fabae*)	Maize + *faba* bean and barley + *faba* bean	Sahile *et al.*, 2008
Beans	Angular leaf spot (*Phaeoisariopsis griseola*)	Maize + bean	Vieira *et al.*, 2009
Pea	Ascochyta blight (*Mycosphaerella pinodes*)	Cereal + pea	Schoeny *et al.*, 2010

There is limited scope of application of chemical herbicides in intercropping systems and chemical weed control (with post emergence herbicides) is not preferred particularly in the mixed stands of dicotyledonous and monocotyledonous crops. In general, intercropping covers maximum ground area; therefore, the weeds population remains less than sole cropping of individual crops. Moreover, root exudates and leaf leachates may exhibit allelopathic effects on weeds that can minimize weed seed bank as well as weed population. Researchers recorded lesser weeds population in different intercropping treatments with maize + legume (Bilalis *et al.*, 2010) and other crop combinations with weeds suppressive nature (Kiroriwal and Yadav, 2013; Gu *et al.*, 2021).

4.4.4 Legume Effect Ensuring Less use of Chemical Fertilizers

The legume crops add nitrogen to the soil through the process of BNF; therefore, presence of legume in an intercropping system is always beneficial to non-legumes. The nitrogen fixed by legumes in mixed stand not only used by them but also, they transfer a portion to non-legumes in association. A study on ^{15}N labeling showed that fixed nitrogen was transferred from soybean to corn in intercropping and seed inoculation of *Glomus mosseae* and *Rhizobium* triggered the process (Ananthi *et al.*, 2017). In a study, Choudhary and Choudhary (2016) recorded that soil phosphorus and potassium balance was enhanced in cereal + legume intercropping system. The enhanced soil fertility directs towards reduced use of chemical nutrients.

4.4.5 Natural Insurance by Crop Diversification

Under risky environmental conditions, crop failure is very common. In tropical climate, impacts of soil moisture stress may be fatal leading to crop failure. In and intercropping system, as two or more crops are grown together, all the crops may not be equally affected by climatic aberrant conditions; thus, some yield may be obtained which can be considered as natural insurance (Maitra *et al.*, 2019). Sometimes, biotic agents can also cause crop failure and because of functional diversity in the intercropping system, all the crops grown in mixed stands are not totally damaged. This unique feature made the intercropping system a popular and effective cropping system for resource-poor smallholders of drylands.

4.4.6 Greater Ecosystem Services

An ecosystem consists of various flora and fauna in the physical environment and interaction of all the components. Under favourable conditions, the biological communities present in the ecology can perform properly and that enhances the ecological soundness. Intercropping system has enough potential to nurture above and below-ground diversity. Further, intercropping causes low carbon emission from agriculture. Earlier research revealed that intercropping combinations of maize + pea and soybean + wheat resulted in less emission of carbon from the field compared to sole cropping of cereals (Chai *et al.*, 2014). Abundance of biological agents is responsible for maintenance of pests population dynamics (Maitra *et al.,* 2021). A healthy ecosystem ensures proper ecosystem services. Amongst various ecosystem services, provisioning good and services is nicely observed with enhanced production of food, forage and feed, biofuel and fuel in intercropping system. Some of the supporting services such as increased pollinator population, carbon sequestration, nutrient cycling and soil health improvement are obtained by adoption of intercropping systems.

The benefits highlighted above are indicative to realizing the holistic advantages of intercropping. Moreover, there is a need for further research to better understand the functioning of intercropping systems targeting a sustainable harvest under the present context of climate change.

References

Adeniyan, O. N., Akande, S. R., Balogun, M. O. and Saka, J. O. 2007. Evaluation of crop yield of African yam bean, maize and kenaf under intercropping systems. *Am. Eurasian J. Agric. Environ. Sci.* **2**(1):99–102.

Adler, P. R., Del Grosso, S. J. and Parton, W. J. 2007. Life cycle assessment of net greenhouse-gas flux for bioenergy cropping systems. *Ecol. Appl.* **17**:675–691.

Ae, N., Arihara, J., Okada, K., Yoshihara, T., Johansen, C. 1990. Phosphorus uptake by pigeon pea and its role in cropping systems of the Indian subcontinent. *Sci.* **248**(4954): 477-480.

Afrin, S., Latif, A., Banu, N. M. A., Kabir, M. M. M., Haque, S. S., Ahmed, M. E. and Ali, M. P. 2017. Intercropping empower reduces insect pests and increases biodiversity in agro-ecosystem. *Agric. Sci.* **8**(10): 1120.

Alvey, S., Yang, C. H., Buerkert, A. and Crowley, D. E. 2003. Cereal/legume rotation effects on rhizosphere bacterial community structure in West African soils. *Biol. Fertil. Soils.* **37**:73–82.

Amanullah, S. K. and Farhan Khalil, I. 2020. Influence of irrigation regimes on competition indexes of winter and summer intercropping system under semi-arid regions of Pakistan. *Sci. Rep.* **10**:8129.

Ananthi, T., Amanullah, M. M. and Al-Tawaha, A. R. M. S. 2017. A review on Maize-Legume intercropping for enhancing the productivity and soil fertility for sustainable agriculture in India. *Adv. Environ. Biol.* **11**(5):49-63.

Assefa, A., Tana, T., Dechassa, N., Dessalgn, Y., Tesfaye, K. and Wortmann, C. S. 2016. Maize - common bean/lupine intercrop productivity and profitability in maize-based cropping system of Northwestern Ethiopia. *Ethiop. J. Sci. Technol.* **9**(2): 69- 85.

Autrique, A. and Potts, M. J. 1987. The influence of mixed cropping on the control of potato bacterial wilt (*Pseudomonas solanacearum*). *Ann Appl Biol.* **111**:125-133.

Bado, B. V., Sedogo, M. P. and Lompo, F. 2004. Long term effects of mineral fertilizers, phosphate rock, dolomite and manure on the characteristics of an ultisol and maize yield in Burkina Faso. In: Bationo A (ed) Managing nutrient cycles to sustain soil fertility in sub-Saharan Africa. Academy Science Publishers, Nairobi, pp. 608.

Barbosa, P. I., Lima, P. S., de Oliveira, O. F. and de Sousa, R. P. 2008. Planting times of cowpea intercropped with corn in the weed control. *Revista. Caatinga*, **21**(1):113-119.

Bilalis, D., Papastylianou, P., Konstantas, A., Patsiali, S., Karkanis, A. and Efthimiadou, A. 2010. Weed-suppressive effects of maize-legume intercropping in organic farming. *Int. J. Pest. Manag.* **56**:173-181.

Buerkert, A., Bationo, A. and Dossa, K. 2000. Mechanisms of residue mulch induced cereal growth increases in West Africa. *Soil Sci. Soc. Am. J.* **64**:346–358.

Burn, A. J., Coaker, T. H. and Jepson, P. C. 1987. Integrated Pest Management. Academic Press, London. p.82.

Chai, Q., Qin, A., Gan, Y. and Yu, A. 2014. Higher yield and lower carbon emission by intercropping maize with rape, pea and wheat in arid irrigation areas. *Agron. Sustain. Dev.* **34**:535-543.

Chen, G., Kong, X., Gan, Y., Zhang, R., Feng, F., Yu, A., Zhao, C., Wan, S. and Chai, Q. 2018. Enhancing the systems productivity and water use efficiency through coordinated soil water sharing and compensation in strip intercropping. *Sci. Rep.* **8**:10494. DOI:10.1038/s41598-018-28612-6.

Chikte, P., Thakare, S. M. and Bhalkare, S. K. 2008. Influence of various cotton-based intercropping systems on population dynamics of thrips, *Scircothrips dorsalis* Hood and whitefly, *Bemisia tabaci* Genn. *Res. Crop.* **9**:683-687.

Choudhary, V. K. and Chaoudhary, B. U. 2016. A staggered maize-legume intercrop arrangement influences yield, weed smothering and nutrient balance in the Eastern Himalayan Region of India. *Exp. Agric.* **54**(2):181-200 DOI: 10.1017/S0014479716000144.

Collins, H. P., Fay, P. A., Kimura, E., Fransen, S. and Himes, A. 2017. Intercropping with switchgrass improves net greenhouse balance in hybrid poplar plantations on a sand soil. Soil. *Sci. Soc. Am. J.* **81**:781-795.

Cu, S. T., Hutson, J. and Schuller, K. A. 2005. Mixed culture of wheat (*Triticum aestivum* L.) with white lupin (*Lupin usalbus* L.) improves the growth and phosphorus nutrition of the wheat. *Plant Soil.* **272**(1- 2):143-151.

Cui, W.T., Gao, C.Y., Fang, P., Lin, G.Q. & Shen, W.B. 2013. Alleviation of cadmium to- xicity in *Medicago sativa* by hydrogen-rich water. *J. Hazard. Mater.* **260**:715-724.

Dahmardeh, M., Ghanbari, A., Syahsar, B.A. and Ramrodi, M. 2010. The role of intercropping maize (*Zea mays* L.) and cowpea (*Vigna unguiculata* L.) on yield and soil chemical properties. *Afr. J. Agric. Res.* **5**(8):631–636.

Dai, J., Qiu, W., Wang, N., Wang, T., Nakanishi, H. and Zuo, Y. 2019. Benefits of Intercropping on Iron Nutrition. *Front. Plant Sci.* **10**:605. Doi: 10.3389/fpls.2019.00605.

Dass, A. and Sudhishir, S. 2010. Intercropping in finger millet (*Eleusine coracana*) with pulses for enhanced productivity, resource conservation and soil fertility in uplands of Southern Orissa. *Ind. J. Agron.* **55**: 89-94.

Dempster, J. P. and Coaker, T. H. 1974. Diversification of crop ecosystems as a means of controlling *pests. In:* Biology of pests and disease control. (*Eds.* Price Jones, D. and Solomon, M. E.) Blackwell, Oxford. pp. 106-114.

Dexter, A. R. 2004. Soil physical quality part I. Theory, effects of soil texture, density, and organic matter, and effects on root growth. *Geoderma.* **120**:201–214.

Epidi, T. T., Bassey, A. E. and Zuofa, K. 2008. Influence of intercrops on pests' populations in upland rice (*Oryza sativa* L.). *Afr. J. Environ. Sci. Technol.* **2**:438-441.

Fester, T., Giebler, J., Wick, L.Y., Schlosser, D. and Kästner, M. 2014. Plant-microbe interactions as drivers of ecosystem functions relevant for the biodegradation of organic contaminants. *Curr. Opin. Biotechnol.* **27**:168–175.

Finckh, M. R., Gacek, E. S., Goyeau, H., Lannou, C., Merz, U., Mundt, C. C., Munk, L., Nadziak, J., Newton, A. C., de Vallavieille-Pope, C. and Wolfe, M. S. 2000. Cereal variety and species mixtures in practice, with emphasis on disease resistance. *Agronomie.* **20**:813-837.

Garg, N. 2007. Symbiotic nitrogen fixation in legume nodules: process and signaling. A review. *Agron. Sustain. Dev.* **27**:59–68.

Ghanbari, A., Dahmardeh, M., Siahsar, B. A. and Ramroudi, M. 2010. Effect of maize (*Zea mays* L.) - cowpea (*Vigna unguiculata* L.) intercropping on light distribution, soil temperature and soil moisture in & environment. *J. Food Agr. Environ.* **8**:102-108.

Gitari, H. I., Nyawade, S. O., Kamau, S., Karanja, N. N., Gachene, C. K. K., Raza, M. A. Maitra, S. and Schulte-Geldermann, E. 2020. Revisiting intercropping indices with respect to potato-legume intercropping systems. *Field Crops Res.* **258**:107957.

Gu, C., Bastiaans, L., Anten, N.P.R., Makowski, D. and van der Werf, W. 2021. Annual intercropping suppresses weeds: A meta-analysis. Agric. Ecosyst. Environ. **322**:107658, https://doi.org/10.1016/j.agee.2021.107658.

Habimana, S., Mbaraka, S. R., Karangwa, A., Rucamumihigo, F. X., Musana, F. R., Mutamuliza, E. and Cyamweshi, A. R. 2021. Effect of intercropping aerobic rice with leafy vegetables on crop growth, yield and its economic efficiency. *Afr. J. Biotechno.* **20**(7), 313-317.

Hauggaard-Nielsen, H., Gooding, M., Ambus, P., Corre-Hellou, G., Crozat, Y., Dahlmann, C., Dibet, A., von Fragstein, P., Pristeri, A., Monti, M. and Jensen, E. S. 2009. Pea–barley intercropping for efficient symbiotic N_2-fixation, soil N acquisition and use of other nutrients in European organic cropping systems. *Field Crops Res.* **113**:64–71. doi:10.1016/j.fcr.2009.04.009.

Hayman, D. S. 1986. Mycorrhizae of nitrogen fixing legumes. *World J. Microbiol. Biotechnol.* **2**(1):121–145.

Hiebsch, C. K. 1978. Interpretation of yields obtained in crop mixture. Abstracts of American Society of Agronomy, Madison, Wisconsin, pp. 41.

Iqbal, N., hussain, S., Ahmed, Z., Yang, F., Wang, X., Liu, W., Yong, T., Du, J., Shu, K., Yang, W. and Liu, J. 2018. Comparative analysis of maize-soybean strip intercropping systems: a review. *Plant Prod. Sci.* **22**(2): 131-142. DOI: 10.1080/1343943X.2018.154137.

Jensen, E. S., Peoples, M. B., Boddey, R. M., Gresshoff, P. M., Hauggaard-Nielsen, H., Alves, B.J. and Morrison, M. J. 2012. Legumes for mitigation of climate change and the provision of feedstock for biofuels and biorefineries. A review. *Agron Sustain Dev.* **32**:329–64.

Jin, Q. J., Zhu, K. K., Cui, W. T., Xie, Y. J., Han, B. and Shen, W.B. 2013. Hydrogen gas acts as a novel bioactive molecule in enhancing plant tolerance to paraquat-induced oxidative stress via the modulation of heme oxygenase-1 signalling system. *Plant Cell Environ.* **36**: 956–969.

Kaiya, S., Utsunomiya, S., Suzuki, S., Yoshida, N., Futamata, H., Yamada, T. and Hiraishi, A. 2012. Isolation and functional gene analyses of aromatic hydrocarbon-degrading bacteria from a polychlorinated dioxin-dechlorinating process. *Microbes. Environ*. **27**: 1112010337-1112010337.

Kermah, M., Franke, A. C., Adjei-Nsiah, S., Ahiabor, B. D. K., Abaidoo, R. C. and Giller, K. E. 2017. Maize- grain legume intercropping for enhanced resource use efficiency and crop productivity in the Guinea of northern Ghana. *Field Crops Res*. **213**:38-50.

Khan, M. N., Muhammad, S., Asrar, M., Ashraf, M. S. and Rafi, Q. 2020. Mungbean (Vigna radiata) intercropping enhances productivity of late season irrigated cotton in Punjab. *Asian J. Agric. Biol*. **8**(4): 472-479.

Kiroriwal, A. and Yadav, R. S. 2013. Effect of intercropping systems on intercrops & weeds 2013. *Int. J. Agric. Food Sci. Technol*. **4**(7):643-646.

Kyamanywa S. and Ampofo, J. K. O. 1988. Effect of cowpea/maize mixed cropping on the incident light at the cowpea canopy and flower thrips (Thysanoptera: *Thripidae*) population density. *Crop Prot*. **7**:186-189.

Lal, R. 2015. Restoring soil quality to mitigate soil degradation. Sustain. **7**:5875–5895. https://doi.org/10.3390/su7055875.

Li, L., Zhang, L. Z. and Zhang, F. Z. 2013. Crop mixtures and the mechanisms of over yielding. *In: Levin SA, ed. Encyclopedia of biodiversity*, 2nd Edn. vol, 2. Waltham, MA, USA: Academic Press, 382–395.

Li, S. and Wu, F. 2018. Diversity and Co-occurrence Patterns of Soil Bacterial and Fungal Communities in Seven Intercropping Systems. *Front. Microbiol*. **9**: 15-21.

Lithourgidis, A. S., Dordas, C. A., Damalas, C. A. and Vlachostergios, D. N. 2011. Annual intercrops: an alternative pathway for sustainable agriculture. *Aust. J. Crop Sci*. **5**:396–410.

Madhu, M., Hombegowda, H. C., Beer, K., Adhikary, P. P., Jakhar, P., Sahoo, D. C., Dash, C. J., Kumar, G. and Naik, G. B. 2019. Status of natural resources and resource conservation technologies in eastern region of India. (*In*: Karma Beer *et al., Eds.*) *Resource conservation in eastern region of India: lead papers of FFCSWR*, 2019, Indian Association of Soil and Water Conservationists, Dehradun, Uttarakhand, pp.60-82.

Mahallati, M. N., Koocheki, A., Mondani, F., Feizi, H. and Amirmoradi S. 2015. Determination of optimal strip width in strip intercropping of maize (*Zea mays* L.) and bean (*Phaseolus vulgaris* L.) in Northeast Iran. *J. Clean Prod*. **106**:343–350.

Maitra S. and Ray, D. P. 2019. Enrichment of biodiversity, influence in microbial population dynamics of soil and nutrient utilization in cereal-legume intercropping systems: A Review. *Int. J. Biores. Sci.* **6**(1):11-19. DOI: 10.30954/2347-9655.01.2019.3.

Maitra, S., Palai J. B., Manasa, P. and Prasanna Kumar D. 2019. Potential of Intercropping System in Sustaining Crop Productivity. *Int. J. Agric. Environ. Biotechnol*. **12**:39-45. DOI: 10.30954/0974-1712.03.2019.7.

Maitra, S., Ghosh, D. C., Sounda, G. and Jana, P.K. 2001. Performance of inter-cropping legumes in fingermillet (*Eleusine coracana*) at varying fertility levels. *Ind. J. Agron.* **46**(1):38-44.

Maitra, S., Ghosh, D. C., Sounda, G. Jana, P. K. and Roy, D. K. 2000. Productivity, competition and economics of intercropping legumes in finger millet (*Eleusine coracana*) at different fertility levels. *Ind. J. Agric. Sci*. **70**(12):824-828.

Maitra, S., Shankar, T. and Banerjee, P. 2020. Potential and Advantages of Maize-Legume Intercropping System (Online First), Intech Open, (In: Maize - Production and Use, Ed. Akbar Hossain), DOI: 10.5772/intechopen.91722. Available from: https://www.intechopen.com (Accessed on 26 March, 2020).

Maitra, S., Hossain, A., Brestic, M., Skalicky, M., Ondrisik, P., Gitari, H., Brahmachari, K., Shankar, T., Bhadra, P., Palai, J.B., Jena, J., Bhattacharya, U., Duvvada, S.K., Lalichetti, S. and Sairam, M. 2021. Intercropping- a low input agricultural strategy for food and environmental security. *Agronomy*. **11**(342):1–28.

Manasa, P., Maitra, S. and Devender Reddy, M. 2018. Effect of Summer Maize-Legume Intercropping System on Growth, Productivity and Competitive Ability of Crops. *Int. J. Manage. Technol. Engg.* **8**(12): 871-2875.

Manasa, P., Maitra, S. and Barman, S. 2020. Yield attributes, yield, competitive ability and economics of summer maize-legume intercropping system. *Int. J. Agric. Environ. Biotechnol.* **13**(1): 3-38, DOI: 10.30954/0974-1712.1.2020.16.

Mandal, M. K., Banerjee, M., Banerjee, H., Alipatra, A. and Malik, G. C. 2014. Productivity of maize (*Zea mays*) based intercropping system during kharif season under red and lateritic tract of West Bengal. *The Bioscan*. **9**(1):31-35.

Meena, V. S., Maurya, B. R., Meena, R. S., Meena, S. K., Singh, N. P. and Malik, V. K. 2014. Microbial dynamics as influenced by concentrate manure and inorganic fertilizer in alluvium soil of Varanasi, India. *Afr. J. Microbiol. Res.* **8**: 257-263.

Mobasser, H. R., Vazirimehr, M. R. and Rigi, K. 2014. Effect of intercropping on resources use, weed management and forage quality. *Int. J. Plant Animal Env. Sci.* **4**(2): 06-713.

Mohammed, E. M., Meawad, A. A., Elesawy, A. E. and Abdelkader, M. A. I. 2022. Maximizing land utilization efficiency and competitive indices of roselle and cluster bean plants by intercropping pattern and foliar spray with Lithovit. *Saudi Journal of Biological Sciences*. https://doi.org/10.1016/j.sjbs.2022.01.026.

Mousavi, S., Yousefi-Moghadam, S., Mostafazadeh-Fard, B., Hemmat, A. and Yazdani, M. R. 2009. Effect of puddling intensity on physical properties of a silty clay soil under laboratory and field conditions. *Paddy Water Environ*. **7**(1):45–5.

Mucheru-Muna, M., Pypers, P., Mugendi, D., Kungu, J., Mugwe, J., Merckx, R. and Vanlauwe, B. 2010. A staggered maize legume intercrop arrangement robustly increases crop yields and economic returns in the highlands of Central Kenya. *Field Crops Res*. **115**:132-139.

Mugwe, J., Mugendi, D., Okoba, B., Tuwei, P. and O'Neill, M. 2004. Soil conservation and fertility improvement using leguminous Shrubs in central highlands of Kenya: NARFP Case study. In: Bationo A (ed.) Managing nutrient cycles to sustain soil fertility in sub-Saharan Africa. Academy Science Publishers, Nairobi, pp. 608.

Narayanaswamy, P., Ganghadharan, K., Chandrasekharan, G., Velazhagan, R. and Karunanidhi, K. 1988. Proceedings of National Workshop on Pests and Diseases, Tamilnadu Agricultural University, 16-18 September.

Nicholls, C. I. and Altieri, M. A. 2013. Plant biodiversity enhances bees and other insect pollinators in agroecosystems: a review. *Agron. Sustain. Develop*. **33**(2):257-274.

Nuruzzaman, M., Lambers, H., Bolland, M. D. A. and Veneklaas, E. J. 2006. Distribution of carboxylates and acid phosphatase and depletion of different phosphorus fractions in the rhizosphere of a cereal and three grain legumes. *Plant Soil* **281**(1):109–120.

Palai, J. B., Malik G. C., Maitra S. and Banerjee, M. 2021. Role of Rhizobium on Growth and Development of Groundnut: A Review. *Int. J. Agric. Environ. Biotechnol.* **14**(1):63-73, DOI: 10.30954/0974-1712.01.2021.7

Panda, S. K., Maitra, S., Panda, P., Shankar, T., Pal, A., Sairam, M. and Praharaj, S. 2021. Productivity and competitive ability of rabi maize and legumes intercropping system. *Crop Res.* **56**:0970-4884.

Peoples, M. B. and Craswell, E. T. 1992. Biological nitrogen fixation: investments, expectations and actual contributions to agriculture. *Plant Soil*. **141**(1-2):13-39.

Peyraud, J. L., Gall, A. L. and Lüscher, A. 2009. Potential food production from forage legume-based systems in Europe: An overview. *Irish. J. Agric. Food Res*. **48**(2):115-135.

Pikul, J. L., Johnson, J. M. F. Schumacher, J. T. E., Vigil, M. and Riedell, W.E. 2008. Change in surface soil carbon under rotated corn in eastern South Dakota. *Soil. Sci. Soc. Am. J.* **72**:1738–1744.

Raji, J. A. 2007. Intercropping soybean and maize in a derived savanna ecology. *Afr. J. Biotechnol.* **6**(16):1885–1887.

Raza, M. A., Cui, L., Qin, R., Yang, F. and Yang, W. 2020. Strip-width determines competitive strengths and grain yields of intercrop species in relay intercropping system. *Sci. Rep.* **10**(1):1-12.

Raza, M.A., Feng L. Y., Van der Wref, W., Iqbal, N., Khan, I., Hassan, M. J., Ansar, M., Chen, Y. K., Xi, Z. J., Shi, J.Y., Ahmed, M., Yang, F. and Yang, W. 2019. Optimum leaf defoliation: a new approach for increasing nutrient uptake and land equivalent ratio of maize soybean relay intercropping system. *Field Crop Res.* **244**:107647. DOI:10.1016/j.fcr.2019.107647.

Ryan, P. R., Tyerman, S. D., Sasaki, T., Furuichi, T., Yamamoto, Y., Zhang, W. H. and Delhaize, E. 2011. The identification of aluminium-resistance genes provides opportunities for enhancing crop production on acid soils. *J. Exp. Bot.* **62**(1):9-20.

Saharan, K., Schütz, L, Kahmen, A., Wiemken, A., Boller, T. and Mathimaran, N. 2018. Finger millet growth and nutrient uptake is improved in intercropping with pigeon pea through "Biofertilization" and "Bioirrigation" mediated by *Arbuscular* mycorrhizal fungi and plant growth promoting rhizobacteria. *Front. Environ. Sci.* **6**(46):1-11.

Sahile, S., Fininsa, C., Sakhuja P. K. and Ahmed, S. 2008. Effect of mixed cropping and fungicides on chocolate spot (*Botrytis fabae*) of faba bean (*Vicia faba*) in Ethiopia. *Crop Protect.* **27**(2): 275-282.

Schädler, S., Morio, M., Bartke, S., Rohr-Zänker, R. and Finkel, M. 2011. Designing sustainable and economically attractive brownfield revitalization options using an integrated assessment model. *J. Environ. Manag.* **92**:827–837.

Schelud'ko, A. V., Makrushin, K. V., Tugarova, A. V., Krestinenko, V. A., Panasenko, V. I., Antonyuk, L. P. & Katsy, E. I.2009. Changes in motility of the rhizobacterium Azospirillum brasilense in the presence of plant lectins. *Microbiol. Res.* **164**:149-156.

Scheublin, T. R., Ridgway, K. P., Young, J. P. W. and Van der Heijden, M. G. A. 2004. Nonlegumes, legumes, and root nodules harbor different arbuscular mycorrhizal fungal communities. *Appl. Environ. Microbiol.* **70**:6240–6246.

Schoeny, A., Jumel, S., Rouault, F., Lemarchand, E. and Tivoli, B. 2010. Effect and underlying mechanisms of pea-cereal intercropping on the epidemic development of *Ascochyta* blight. *Eur. J. Plant Pathol.* **126**:317-331.

Signor, D. and Cerri, C. P. E. 2013. Nitrous oxide emission in agricultural soils: a review. *Pesq. Agropec. Trop. Goiânia.* **43**:322–338.

Singh, D., Mathimaran, N., Boller, T. and Kahmen, A. 2020. Deep-rooted pigeon pea promotes the water relations and survival of shallow-rooted finger millet during drought—Despite strong competitive interactions at ambient water availability. PLoS ONE. **15**(2):e0228993. https://doi.org/10.1371/journal.pone.0228993.

Song, Y. N., Zhang, F. S., Marschner, P., Fan, F. L., Gao, H. M., Bao, X. G., Sun J. H. and Li, L. 2007. Effect of intercropping on crop yield and chemical and microbiological properties in rhizosphere of wheat (*Triticum aestivum* L.), maize (*Zea mays* L.), and faba bean (*Vicia faba* L.). *Biol. Fert. Soils.* **43**(5):565-574.

Spehn, E. M., Joshi J., Schmid, B., Alphei, J. and Korner, C. 2000. Plant diversity effects on soil heterotrophic activity in experimental grassland ecosystems. *Plant Soil.* **224**:217–230.

Srinivasarao, C., Venkateswarlu, B. and Lal, R. 2012. Long-term effects of soil fertility management on carbon sequestration in a rice-lentil cropping system of the Indo-Gangetic plains. *Soil Sci. Soc. Am. J.* **76**(1):167–178.

Suman, A., Lal, M., Singh, A.K. and Gaur, A. 2006. Microbial biomass turnover in Indian subtropical soils under different sugarcane intercropping systems. *Agron. J.* **98**(3):698–704.

Turnbull, J. and Bowman, W. D. 2002. Variable effects of nitrogen additions on the stability and turnover of soil carbon. *Nature.* **419**:915–917.

Vieira, R. F., de Paula Junior, T. J., Teixeira, H. and Vieira, C. 2009. Intensity of angular leaf spot and anthracnose on pods of common beans cultivated in three cropping systems. *Ciénc Agrotec.* **33**:1931-1934.

Willey, R. W. 1979. Intercropping—Its importance and research needs. Part 1: Competition and yield advantages. *Field Crops Res*. **32**:1-10.

Willey, R. W., and Osiru, D. S. O. 1972. Studies on mixtures of maize &and beans (*Phasrolus vulgaris*) with particular reference to plant population. *J. Agri. Sci. Camb*. **79**:519-529.

Xue Y., Xia, H., Christie, P., Zhang, Z., Li, L. and Tang, C. 2016. Crop acquisition of phosphorus, iron and zinc from soil in cereal/legume intercropping systems: a critical review. Ann. Bot. Lond. **117**:363–77. doi:10.1093/aob/mcv182.

Zhang, F., Shen, J., Li, L. and Liu, X. 2004. An overview of rhizosphere processes related with plant nutrition in major cropping systems in China. *Plant Soil*. **260**(1- 2):89-99.

5

Organic Agriculture and Intercropping System

5.1 Modern Agriculture: Background and Present Issues in India

Agriculture is an important factor for human civilization as it not only provides food, feed for the livestock, fibre, fuel and industrial raw material but also supports the livelihood of a wide range of world population. Besides, it shows its prominent presence in the economy of the developing countries. Being a developing country, agriculture is the largest sector in India where about 54% of the population is engaged. The growth of agriculture in India during last seven and half decades (after independence) is quite remarkable in terms of sufficient food-grains production as well as in achieving self-sufficiency in food production. The task to achieve it was not so easy.

In the British ruling, our traditional farming was mostly dependent on the blessing of monsoon and it was mainly handled by the small and marginal farmers who used to produce the crop for subsistence of their family and sell the surplus produce to the local market. During that period, farmers generally used to adopt crop management practices such as providing nutrients, maintenance of soil fertility and control of pests, diseases and weeds in their own indigenous ways and these were done without application of chemical inputs. Undoubtedly, these inputs were made up of locally available ingredients, farm and animal wastes, plant origin and those too of low cost. The incidence like crop failure was there due to drought and flood which led to famine also. The British revenue system and in the colonial India as well as feudalistic pressure of landlords compelled the small and marginal farmers even to revolt against the system. But during the post-independence era, more especially in 1950s onwards mitigation of hunger was a great challenge in India and our policy makers had no options rather to welcome the 'Green Revolution'. The Indian Government selected the then Punjab (Punjab and Haryana of the present) as the suitable area for trying the new technologies because of available water resources and assured irrigation. India started its own 'Green Revolution' by initiation of crop improvement, creation of irrigation potential and promotion of chemical agro-inputs. By adoption of Green Revolution Technologies (GRTs), India succeeded in increasing grain production.

The country's image of a 'begging bowl' dramatically changed into self-sufficiency by adopting the Green Revolution over the next two decades. The improved technologies like adoption of high yielding varieties, providing chemical fertilizers for supply of nutrients, chemical pesticides for plant protection and creation of irrigation source and providing water to crops were the main inputs which led to increase in yield. Considering the situation of that time, it may be commented that welcoming the green revolution was the correct one. Later, the country experienced with white, blue and yellow revolutions as milestones of Indian agriculture, but all these achievements were based on adoption of inputs supply driven technologies.

Actually, during post Green Revolution period, the production sustainability appeared into big threat due to non-judicious use of high value chemical inputs, luxurious use of irrigation water, genetic erosion and mono-cropping, land degradation etc. and these ultimately showed a stagnation or declining trend in productivity and vulnerability of agro-ecosystem. The global warming as well as climate change put an added fuel to this crisis which compelled technocrats and environmentalists to think of alternatives. There is a need for increased food production under the consequences of dwindling land and water resources and climatic aberrations. The "seed" of organic farming is actually sown in the basement of some alternative thoughts. There is need to encourage a sound, cautiously viable, environmentally non-degrading, and socially acceptable use of all natural resources such as land, water and genetic legacy towards assured agricultural sustainability.

On the other hand, the green revolution-based supply driven technologies and later modern industrialized agriculture played a vital role in degrading environment and agroecosystem, contributing GHGs emission and thus, triggering climate change, polluting freshwater and ecology, and unflattering soil fertility as anthropogenic interventions. All the practices further led to ignoring natural resources with havoc and unjustified use of agrochemicals for supplying plant nutrients, protecting plants and boosting the crop growth. Agriculture has a direct relationship to the natural resources and because of the detrimental impacts of modern agriculture, the agroecosystem comprising of natural resources is also in dwindling condition. Thus, modern agriculture is unable to confirm agricultural sustainability. The yield plateauing was recorded in the 'so-called' promising food-baskets of the country because of over-exploitation and natural resources and adoption of defective farming technologies declining agroecosystem.

In the context of achieving agricultural sustainability, agriculture should be directed towards ecofriendly way as it can fulfill the demand for the present as well as for the future. food and nutritional security demands access to enough and nutritious foods for all and hence, there is an urgent requirement for development of such an agricultural practice that will not harm the agroecosystem. In this regard organic farming or organic agriculture can be thought of as it has no negative impacts. It is such a production system that is devoid of synthetically produced agrochemicals and inputs, and therefore, there is least possibility of degradation of agroecosystem.

In the recent period, growth of organic food market in some developed western countries that are scared of consuming the foods grown with chemical fertilizers and pesticides residue has given further impetus to the promoters of organic agriculture in India and thus, a potential horizon of exporting organic agriculture products has been opened leading to the opportunity to earn foreign currency. During last two decades organic farming has gained considerable respect and acceptance in our country. At present, there are several farmers who are producing organically raised crops, particularly cereals, vegetable, fruits etc. for the elite and health-conscious customers who are having higher purchasing capacity. Thus, the growing demand of organic agricultural products has driven the farmers to adopt organic agriculture in India. Government of India has also structured organic agriculture in the country through adoption of National Programme for Organic Production (NPOP) in April 2000. Presently, India ranks first in the world in terms of number of certified organic growers and eighth in land engaged in organic agriculture (APEDA, 2021). Also, the country has 43.4 lakh ha of certified organic farmland under NPOP inclusive of harvests from 16.8 lakh ha of wild collection. In the year 2020-21 it produced about 35 lakh MT of certified organic products and earned about 1041 million US$ from export (APEDA, 2021).

5.2 Organic Agriculture: A Way Forward

Before the beginning of the twenty-first century, organic agriculture was well defined and structured by several countries. Different leading organizations define organic agriculture and others almost followed the same. In table 5.1, some definitions have been presented.

Table 5.1. Definition of organic agriculture

Defined by	Definition	Year	References
USDA National Organic Standards Board (NOSB)	Organic agriculture is an ecological production management system that promotes and enhances biodiversity, biological cycles and soil biological activity. It is based on minimal use of off-farm inputs and on management practices that restore, maintain and enhance ecological harmony. Organic agriculture practices cannot ensure that products are completely free of residues; however, methods are used to minimize pollution from air, soil and water.	1995	USDA, 2021

IFOAM General Assembly	Organic agriculture is a production system that sustains the health of soils, ecosystems, and people. It relies on ecological processes, biodiversity and cycles adapted to local conditions, rather than the use of inputs with adverse effects. Organic Agriculture combines tradition, innovation, and science to benefit the shared environment and promote fair relationships and good quality of life for all involved.	2008	IFOAM, 2021
FAO/WHO Codex Alimentarius Commission	Organic agriculture is a holistic production management system which promotes and enhances agroecosystem health, including biodiversity, biological cycles, and soil biological activity. It emphasizes the use of management practices in preference to the use of off-farm inputs, taking into account that regional conditions require locally adapted systems. This is accomplished by using where possible, agronomic, biological, and mechanical methods, as opposed to using synthetic materials, to fulfil any specific function within the system.	1999	FAO, 2021

The definition of organic agriculture clearly reveals that it is an agricultural system that leads towards healthy soil, ecosystem services and benefits of the people. It 'relies' on various processes related to crop ecology and maintenance of biodiversity, rather than dumping of synthetically produced chemical inputs showing negative impacts on agroecosystem. In other words, it may be said that organic agriculture is comprised of traditional farming practices, innovation and scientific approaches to facilitate the shared environment, encourage fair relationship and improve the quality of lifeforms. Further, to realize and handle ecological, social and economic apprehensions collectively organic agriculture can be considered as an active approach. In some countries like India and others which nurtured ancient civilizations, organic agriculture was practiced traditionally and India's agrarian history and heritage indicated the long tradition of organic agriculture for the millenniums. Since the early Vedic civilization, agricultural practices such as crop rotation, use of organic inputs, inclusion of crop residues to farm land were evidenced which were actually different forms of soil health and crop nutrient management with locally available inputs and practices. Today's organic agriculture also accentuates use of on-farm inputs based on local availability in a closed system. To understand the philosophy of organic agriculture, it is necessary to look into its principles as mentioned by the International Federation on organic agriculture movement (IFOAM) and presented in Figure 5.1. IFOAM mentioned "these principles are the roots from which organic agriculture grows and develops. They express the contribution that organic agriculture can make to the world, and a vision to improve all agriculture in a global context."

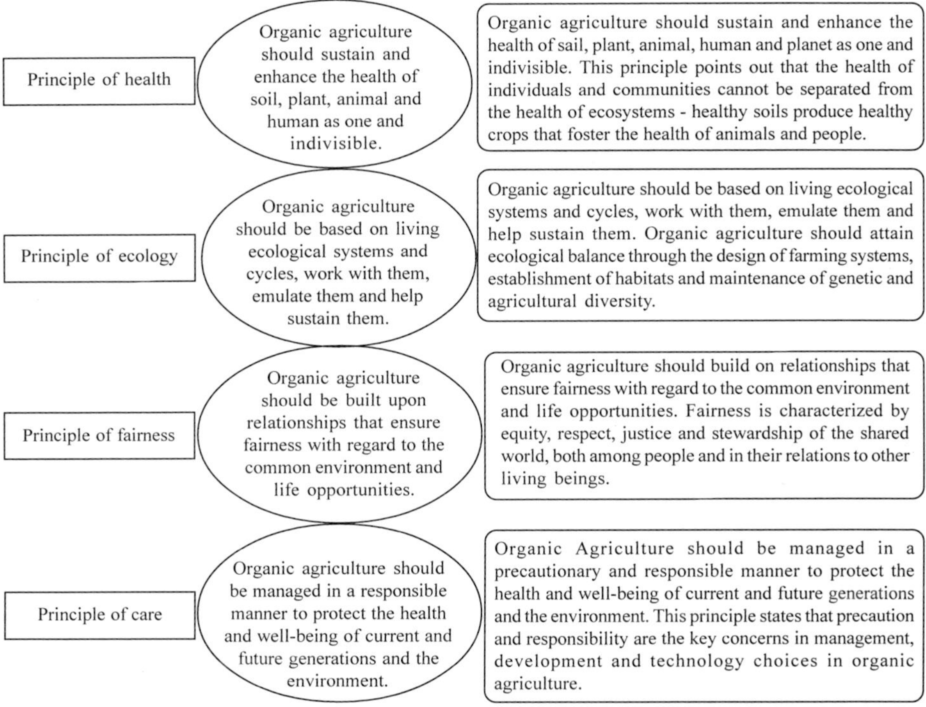

Figure 5.1. Principles of organic farming

In India, agriculture is a tradition and before the green revolution, the country's agriculture was predominantly organic inputs based. During present times also, small and marginal farmers adopt an integrated farming system approach with a combination of crops and allied activities where use of on-farm inputs is common. But, in the original green revolution tract and some other parts of India, modern agricultural practices were adopted with higher investment and use of synthetic agrochemicals. Initially, modern and industrialized agriculture helped in boosting production and productivity of crops, but over years the ill effects were recognized. Now, to fulfill the needs of the present and anticipated future population, the country needs to produce more in a sustainable manner. This requirement has changed the direction of crop production towards using more organic inputs.

Simultaneously, worldwide development and structurization of organic agriculture during two decades back have opened another potential window for Indian farmers to earn foreign currency in the form of export. The National Program for Organic Production (NPOP) was initiated in India in 2000 where the guidelines for organic agriculture were prepared in parity with the standards suggested by CODEX Alimentarius [United Nations Food and Agriculture Organization (FAO) and the World Health Organization (WHO)] and IFOAM. The standards maintained by NPOP are recognized by different developed countries and organic agricultural produces are of satisfactory demand in the export market.

As in organic agriculture, synthetic and outsourced nutrients and other inputs are restricted, the farming system plays an important role. Intercropping is considered in organic agriculture as a holistic approach for maintaining soil fertility, utilizing resources efficiently, enhancing biodiversity and facilitating better ecosystem services.

5.3 Organic Agriculture and Intercropping

The success or failure of organic agriculture greatly depends on the management practices adopted in which crop rotation and diversification, cropping and farming system, soil fertility management and plant protection are important. All of the mentioned qualitative and quantitative aspects can be fulfilled by adoption of suitable intercropping systems (Maitra and Gitari, 2020). Continuous improvement of soil fertility is important in organic agriculture and in this regard, the intercropping system is of prime importance in improvement of soil health. Soil health denotes physical, chemical and biological properties favourable for plant growth. These properties are influenced by external factors such as tillage, crop rotation, cropping system, climate and nutrients application. Intercropping system alone can contribute to soil health improvement.

In general, the crop-livestock farming system is adopted in organic agriculture where the byproducts of livestock are used to crop fields for maintenance and improvement of soil fertility as well as soil health. But where livestock is not used in a farming system or where only crops are raised by following appropriate cropping system in organic agriculture, the intercropping system with legume as a component plays a vital role. Legume has the quality of biological nitrogen fixation (BNF) and keeps its presence in the nitrogen economy of the cropping system (Amosse´ *et al*., 2013). The nitrogen fixed by the legumes in a combination with non-legumes, is utilized by not only legumes but also non-legume components in mixed stands get the benefit. In an intercropping system, legumes are known to share N with non-legumes and the nitrogen transfer ultimately benefits the crops grown in the mixed stand. The biologically fixed nitrogen transfer from legumes to non-legumes in an intercropping system may take place by various ways such as senescence of roots and nodules, root exudates, leaf fall and N-leaching from legume leaves (Fujita *et al*., 1992; Ledgard and Giller, 1995). The quantity of fixed and transferred nitrogen in a legume and non-legume intercropping system varies based on the legume crops selected, their duration, morphology and management practices adopted. The long duration pigeon pea fixes more nitrogen biologically than a short duration legume such as green gram or black gram. The experimental evidence showed that inoculation of *Rhizobium* SH212 strain along with *Glomus mosseae* in an intercropping system with maize + soybean increased nitrogen transfer to non- legumes as recorded by ^{15}N labeling (Senaratne *et al.*, 1995). However, the nitrogen fixed biologically by legumes is not only transferred to component non-legume crops and utilized by them in the intercropping system but also succeeding crops grown in the sequential cropping get the benefit of the leftover N. Further, Xue *et al.* (2016) recorded the phosphorus, iron and zinc nutrition in cereal + legume intercropping system as they have noted and increased availability of these in intercropping. Relay intercropping in cereal-based cropping system is also known to enrich chemical composition of soil as well as soil fertility as the legumes perform the BNF (Amosse´ *et al.*, 2013).

The combined biomass yield in the intercropping system achieved more than sole cropping from a unit area as the crops together utilize more of available resources (namely, carbon dioxide, sunlight, nutrients, soil moisture etc.). Use of more carbon dioxide from the atmosphere is synonymous to reduction of atmospheric pollution. In this regard, it should be mentioned that an intercropping system sequesters more carbon than sole cropping (Cong et al., 2015). Increase in soil organic carbon (SOC) is related to enhanced soil fertility and enhancement of SOC has another importance in terms of mitigation of atmospheric carbon dioxide due to anthropogenic interference (Lal, 2004). As intercropping increases more biomass from a unit area, it is expected that more litter input into the soil which results in carbon sequestration (Ghosh *et al.,* 2006; Li *et al.,* 2011). Somehow, it is matching to the philosophy and principles of organic agriculture and thus, the intercropping system can be considered as an integral part of organic agriculture.

In an intercropping system, solar energy is efficiently used by the crops in a mixture for production of crop yields and biomass. Harvesting of more renewable solar energy for biomass production without any harm to the agroecosystem is a positive phenomenon in terms of production of food, feed and fuel for mankind. Further, morphologically different crop species are preferably chosen in an intercropping system and a combination of deep and shallow-rooted crops is always preferable to ensure better utilization of inherent soil nutrients. Such combination of crops also uses soil moisture available in different layers. Bio-irrigation is another factor which is important in realizing soil moisture availability to shallow rooted crop (finger millet) in combination with deep rooted crop (red gram) in an intercropping in presence of biofertilizers inoculation

(Saharan *et al.*, 2018). Research carried out in the direction of soil moisture availability to crops in intercropping showed that mixed stands of maize + legume reduced soil moisture evaporation and enhanced yield and water use efficiency (Rahman *et al.*, 2017).

In organic agriculture, qualitative improvement of soil is more important and there is limited (or restricted) scope for soil amendments for fertility enrichment. Therefore, loss of top soil as well as nutrients included in loss of topsoil is to be checked by management practices. Intercropping plays an important role in this aspect as it covers more ground area in mixed stand and hence, in erosive soils, it controls runoff of water, soil and nutrients loss in the form of soil loss (Dass and Sudhishir, 2010). In sloppy lands, an intercropping combination of finger millet and black gram was recognized to minimize the runoff, and loss of soil nutrients as the legume component covered more ground area (Dass and Sudhishir, 2010). The study revealed that the combination of finger millet and legumes in erosion prone areas enhanced soil fertility by litters and BNF of legumes and it further resulted in reduced soil and nutrient loss. Moreover, legumes can be recognized as an energy source when digested in the bio-gas chamber and it yields renewable energy with an additional advantage of slurry as organic manure (Maitra and Gitari, 2020). As the chemical nitrogenous fertilizer production procedure is harmful to environment, soil fertility enhancement by addition of slurry is beneficial in organic agriculture in increasing soil fertility and supplying of nutrients to crops. In this way, the intercropping system is considered as an environment-friendly approach in organic agriculture.

Intercropping system is itself a diversified form of agriculture where above- and below-ground diversity are observed (Scherr and McNeely, 2008). The diversity is not only created by cultivating more than one crop in mixture but also an enhanced population of different natural enemies and pollinators adding value (Nicholls and Altieri, 2013; Maitra and Ray, 2019). Earlier researches showed that intercropping flowering plants in maize enhanced biodiversity of pollinators (Schulz *et al.*, 2020). Amy *et al.* (2018) studied on strip cropping of flowers in wheat and noted a greater diversity of pollinators and natural enemies. Norris *et al.* (2018) observed an enhancement in diversity and density of pollinators when flowering plants were intercropped in maize. Gowton *et al.* (2021) mentioned that the plants that produce high level of volatile organic compounds can confuse the harmful insects and they recorded that intercropping peppermint with ryegrass and clover mixes reduced the population of *Drosophila suzukii* with an abundance of pollinators. Intercropping combination with legumes is known to host predators that help to maintain the pest population in the crop field (Maitra and Gitari, 2020). Enrichment of biodiversity helps in ensuring greater ecosystem services and agricultural sustainability which covers the principles of organic agriculture.

In the case of weed management practices in organic agriculture, chemical herbicides are not used and focus is given to mechanical, cultural and biological methods. As in the intercropping system, more ground area is covered and weed infestation remains less, for weed management also the intercropping system can be considered as an ideal option in organic agriculture.

The below-ground diversity is also enriched in the intercropping system in the form of enrichment of the microbial population. Organic manures and crop residues are incorporated in organic agriculture and those also increase microbial population in soil. Legumes are known to ensure enrichment of population of soil microbes such as *Alphaproteo bacteria, Betaproteo bacteria* and *Cyano bacteria* (Li and Wu, 2018). *Rhizobium* is associated with legumes symbiotically and hence, the legumes in intercropping combination increases beneficial microorganisms in the soil. Moreover, in *Rhizobium*-rich soils, population of other helpful microorganisms such as *Arthrobacter simplex*, *Bacillus amyloliquefaciens*, *B. laevolacticus*, *Pseudomonas rathonis*, *P. denitrificans* are increased in the rhizosphere. *Rhizobium* also reduces the population pathogens such as *Fusarium* spp. and *Phytophthora* spp. (Maitra and Ray, 2019). Therefore, microbial diversity is favourable to a healthy agroecosystem which is ultimately reflected in agricultural sustainability.

One of the important benefits of intercropping is creation of above- and below-ground diversity. Creation of biodiversity is essential in organic agriculture. Thus, adoption of intercropping (preferably with legumes) in organic agriculture can be more focused not only for harnessing other benefits but also for diversity in agroecosystem and agricultural sustainability.

References

Amosse´, C., Jeuffroy, M., Mary, B. and David, C.2013. Contribution of relay intercropping with legume cover crops on nitrogen dynamics in organic grain systems. *Nutr. Cycl. Agroecosys.* **98**(1): 1-14, DOI 10.1007/s10705-013-9591-8.

Amy, C., Noël, G., Hatt, S., Uyttenbroeck, R., Van de Meutter, F., Genoud, D., and Francis, F. (2018). Flower strips in wheat intercropping system: Effect on pollinator abundance and diversity in Belgium. *Insects*, **9**(3): 114. https://doi.org/10.3390/insects9030114.

APEDA, 2021. Organic Products, National programme for organic production (NPOP), http://apeda.gov.in/apedawebsite/organic/Organic_Products.htm (Accessed 25 December, 2021).

Cong, W-F., Hoffland, E., Li, L., Six, J., Sun, J-H., Bao, X-G., Zhang, F-S., Van Der Werf, W. 2015. Intercropping enhances soil carbon and nitrogen. *Global Change Biol.*, **21**(4), 1715–1726. doi:10.1111/gcb.12738.

Dass, A. and Sudhishir, S. 2010. Intercropping in finger millet (*Eleusine coracana*) with pulses for enhanced productivity, resource conservation and soil fertility in uplands of Southern Orissa. *Ind. J. Agron.*, **55** (2):89-94.

FAO. 2021. Food and Agriculture Organization, What is organic agriculture? https://www.fao.org/organicag/oa-faq/oa-faq1/en/ (Accessed 25 December 2021).

Fujita, K., Ofosu-Budu, K. G. and Ogata, S. 1992. Biological nitrogen fixation in mixed legume-cereal cropping systems. *Plant Soil*. **141**:155-176.

Ghosh, P.K., Manna, M.C., Bandyopadhyay, K.K., Tripathi, A.K., Wanjari, R.H., Hati, K.M., Misra, A.K., Acharya, C.L. and Subba Rao, A. 2006. Interspecific interaction and nutrient use in soybean/sorghum intercropping system. *Agron. J.* **98**(4): 1097-1108.

Gowton CM, Cabra-Arias C and Carrillo J (2021) Intercropping with peppermint increases ground dwelling insect and pollinator abundance and decreases *Drosophila suzukii* in fruit. *Front. Sustain. Food Syst.* **5**:700842. doi: 10.3389/fsufs.2021.700842.

IFOAM, 2021. Organic agriculture. https://www.ifoam.bio/why-organic/organic- landmarks/definition-organic. (Accessed 25 December, 2021)

Lal, R. 2004. Soil carbon sequestration impacts on global climate change and food security. *Sci.* **304**:1623–1627.

Ledgard, S.J. and Giller, K.E. 1995. Atmospheric N_2 fixation as Alternative Nitrogen Source. *In:* Bacon, P. (*Ed.*) Nitrogen Fertilization and the Environment. Marcel Dekker, New York, pp. 443–486.

Li, L., Sun, J.H., Zhang, F.S. 2011. Intercropping with wheat leads to greater root weight density and larger below-ground space of irrigated maize at late growth stages. *Soil Sci. Plant Nutri.* **57**:61–67.

Li, S. and Wu, F. 2018. Diversity and co-occurrence patterns of soil bacterial and fungal communities in seven intercropping systems. *Front. Microbiol.* **9:**15-21. DOI:10.3389/fmicb.2018.01521.

Maitra, S. and Gitari, H.I. 2020. Scope for Adoption of Intercropping System in Organic Agriculture. *Ind. J. Nat. Sci.* **11**(63): 28624-28631.

Maitra, S. and Roy, D.P. 2019. Enrichment of biodiversity, influence in microbial population dynamics of soil and nutrient utilization in cereal-legume intercropping systems: A review. *Int. J. Biores. Sci.* **6**(1):11-19. DOI: 10.30954/2347-9655.01.2019.3.

Nicholls, C.I. and Altieri, M.A. 2013. Plant biodiversity enhances bees and other insect pollinators in agro-ecosystems. A review. *Agron. Sust. Dev.* **33**(2):257-274. DOI:10.1007/s13593-012-0092.

Norris, S. L., Blackshaw, R. P., Critchley, C. N. R., Dunn, R. M., Smith, K. E., Williams, J., Randall, N. P. and Murray, P. J. 2018. Intercropping flowering plants in maize systems increases pollinator diversity. *Agric. Forest Entomol.* **20** (2):246-254. https://doi.org/10.1111/afe.12251.

Rahman T, Liu X, Hussain S, Ahmed S, Chen G, Yang F, Lilian Chen, L., Du, J., Liu1, W. and Yang, W. 2017 Water use efficiency and evapotranspiration in maize-soybean relay strip intercrop systems as affected by planting geometries. *PLoS ONE* **12**(6):e0178332. https:// doi.org/10.1371/journal.pone.0178332.

Saharan K, Schütz L, Kahmen A, Wiemken A, Boller T and Mathimaran N. 2018. Finger millet growth and nutrient uptake is improved in intercropping with pigeon pea through "Biofertilization" and "Bioirrigation" mediated by arbuscular mycorrhizal fungi and plant growth promoting Rhizobacteria. *Front. Environ. Sci.* **6**:46. doi: 10.3389/fenvs.2018.00046.

Scherr, S.J. and McNeely, J.A. 2008. Biodiversity conservation and agricultural sustainability: towards a new paradigm of 'eco-agriculture' landscapes. *Philos Trans Royal Soc*. **363**:477-494.

Schulz, V.S., Schumann, C., Weisenburger, S., Müller-Lindenlauf, M., Stolzenburg, K. and Möller, K. 2020. Row-intercropping maize (*Zea mays* L.) with biodiversity-enhancing flowering-partners-effect on plant growth, silage yield, and composition of harvest material. *Agric*., **10**: 524; doi:10.3390/agriculture10110524.

Senaratne, R., Liyanage, N. and Soper, R. J. 1995. Nitrogen fixation of and N transfer from cowpea, mungbean and groundnut when intercropped with maize. *Nutr. Cycl. Agro Ecosyst.* **40**(1):4148.

USDA, 2021. U. S. Department of Agriculture, Organic production/organic food: Information access tools, www.nal.usda.gov/legacy/afsic/organic-production organic-food- information-access-tools (Accessed 25 December, 2021).

Xue, Y., Xia, H., Christie, P., Zhang, Z., Li, L. and Tang, C. 2016. Crop acquisition of phosphorus, iron and zinc from soil in cereal/legume intercropping systems: a critical review. *Ann Bot.* **117**:363–377.

6

Intercropping System and Agricultural Sustainability

6.1 Modern Agriculture and Cropping System

Worldwide, agriculture contributes a lot for mankind by providing food and feed for human beings and livestock, respectively. The countries where agriculture plays a vital role, it has enough importance in the economy of the country as well as gross domestic product (GDP). India is an agriculture-based country which nurtures more than half of its population. Under the pressure of continuous growth of human population, there was no option and India adopted modern agricultural technologies which were basically inputs supply-driven approaches and a typical simplification in farming practices. The situation was not different for other developing and so-called agriculture-based countries in the world.

The success of agriculture production actually lies on the interaction of various factors. There is no doubt that in the process of increasing agricultural production, inputs play an important role; but these are not the only factors. The present context is a little bit complicated where declining of agriculture land and natural resources have been prominently noticed. Modern and industrialized agriculture ensured productivity enhancement in short-term basis with use of synthetic and chemical inputs accompanied with provision of irrigation water. Unfortunately, all these created negative impact on agroecosystem, human and animal health, and the quality of agricultural products. Moreover, modern technologies driven monoculture enhanced farm productivity, but at a cost of environmental degradation.

Under the present situation, there is a tremendous pressure for enhancement of farm productivity. But, unless degradation of natural resources is checked, it is very difficult to achieve agricultural sustainability. In contrast to modern agriculture, agricultural sustainability focuses on the holistic approach where diversity is nurtured properly targeting fulfillment of the needs for the present and future. Creation of diversity in agriculture and managing it effectively for agricultural sustainability are vital under the limited resource-base. Adoption of appropriate crops and cropping systems are the only option for the goal of sustainability.

A system is composed of various components which are inter-related. Emerson (2007) mentioned that in cropping system crops are grown in sequential order or in mixed stands and their spatial and temporal interaction with the available resources. In general, farmers aim in achieving more yield and economic return form a unit area (Hauggaard-Nieson, 2001) along with uninterrupted production. The system approach, thus, utilizes the available resources for better farm-output. In this regard, intensification of cropping is one of the meaningful targets. For crop intensification, two options are there; these are either sequential cropping or intercropping. The sequential cropping is widely adopted because it is a simple version of cropping system where all technologies of modern agriculture can be adopted in sequence and periodically. However, intercropping measures the efficiency of cropping systems more accurately in terms of space and time dimensions (Willey *et al.*, 1983). The negative impacts of modern agriculture such as vulnerability in agroecosystem, overexploitation of natural resources and use of huge synthetic chemicals are not common in intercropping (Jackson *et al.*, 2007; Scherr and McNeely, 2008). Thus, intercropping system may be considered as sustainable practice of farming (Tilman *et al.*, 2002; Lichtfouse *et al.*, 2009). The dimension of intercropping is vast and it can be considered as a combination of many systems targeting future needs for the foods and nutritional security. Further, intercropping system is more acceptable to the small holders in the developing countries where most of the farmers belong to this category and intercropping system not only ensures food and nutritional security but also offers more farm output from unit area and yield stability. The yield stability is resultant of species diversity and enhancement of soil quality (Li *et al.*, 2001). Considering all the aspects, intercropping can be evaluated for agricultural sustainability.

6.2 Intercropping and Sustainable Intensification of Cropping System

The modern technologies of crop intensification simplified the cropping system by exogenous application of synthetic chemical inputs and mechanization in sequential cropping. Such development initially resulted in crop productivity enhancement. But the limitations were overdependence on off-farm input supply and sometimes non-judicious use of chemicals, overexploitation of natural resources & excessive energy use. Further, the importance of crop rotation and diversification was also ignored. In this way, the functionality of the agroecosystem was totally neglected which negatively impacted on the cropping system.

The negative impacts of modern agriculture were reflected worldwide, but the developed nations switched to the good agricultural practices (GAP) quickly.

The problem appeared more severe to the developing nations where demand for food is more because of population, poor economic growth of the countries and least investment in agriculture. Modern agricultural technologies resulted in inefficient use of natural resources and caused a massive carbon footprint in agriculture (Lal, 2004) and the agriculture-based developing countries are suffering most. The expansion of population growth demands more farm output. An estimate showed that in the middle of the current century, food and fresh water demand will be 70-85% and 30-85% more than present level, respectively (McMichael *et al.*, 2007). This indicates the need for crop intensification, but it should be in a sustainable way without damaging the agroecosystem and confirming present and future demand.

Intercropping is such a cropping system where all the beneficial practices are adopted inclusive of efficient use of natural resources (land, water, sunlight etc.) and reduction of off-farm inputs and energy. These were ultimately reflected into sustainability of crop intensification (Figure 6.1). The outcomes of sustainable crop intensification are enhanced agroecosystem diversity and soil health improvement, increased crop productivity and yield stability. Moreover, it ensures greater agroecosystem services which is more required in agricultural sustainability (Maitra *et al.*, 2020).

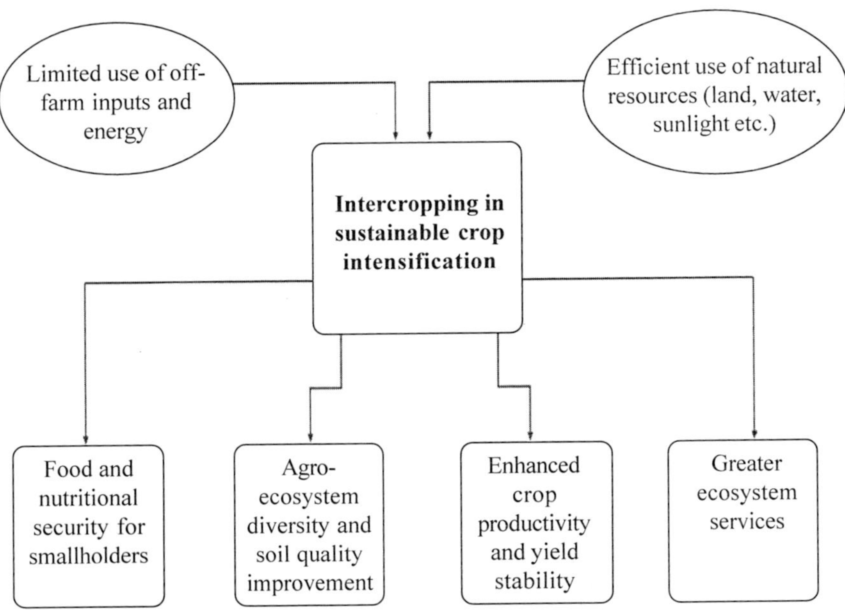

Figure 6.1. Intercropping in sustainable crop intensification

In the developing countries, investment in agriculture is less and a major portion of the population is engaged in agriculture. In this condition, efficient utilization of available resources is to be considered with a priority in utilization of partially engaged family labourers. Food and nutritional security and livelihood improvement of smallholders in the developing countries can be ensured by adoption of an intercropping system. Intercropping is a low-input and energy-efficient cropping system (Maitra *et al.*, 2019). Biodiversity in agricultural systems can be enriched by spatial and temporal intensification of crops (Altieri, 1999). Abundance of above and below ground fauna further adds the value of biodiversity (Maitra and Ray, 2019). Intercropping offers stimulating nature's principles of diversity on farm lands. An appropriate mix of crops and their proportion (inclusive of legumes as component) ensure multiple benefits aiming agricultural sustainability.

Ecosystem services are essential in agricultural sustainability and biodiversity in ecosystems is related to ecosystem services. During the last seven decades or more the ecosystem was changed a lot by anthropogenic intervention for achieving the short-term goals. As a result, erosion of biodiversity and ever exploitation as well as degradation of natural resources were caused. Further, it led to reduction of ecological resilience and disruption of ecosystem services (Jackson *et al.*, 2007). There are four ecosystem services, namely, provisioning, regulating, cultural, and supporting (Maitra *et al.*, 2020). Intercropping is totally supporting to the ecosystem services (Maitra *et al.*, 2021) and it can also be relooked into in the arena of ecosystem services (Figure 6.2). Intercropping provides sufficient food, animal feed and fodder and fuel. It creates functional diversity, manages pests-disease population dynamics, and facilitates nutrient cycling and carbon sequestration. Under soil moisture stress and high temperature conditions, soil evaporation loss is reduced in intercropping and crops get a soothing micro-climate at canopy level. Further bio-irrigation is another beneficial phenomenon under soil moisture stress conditions. Moreover, intercropping assures food and nutritional security to smallholders, enhances yield from unit area, increases income and supports human well-being.

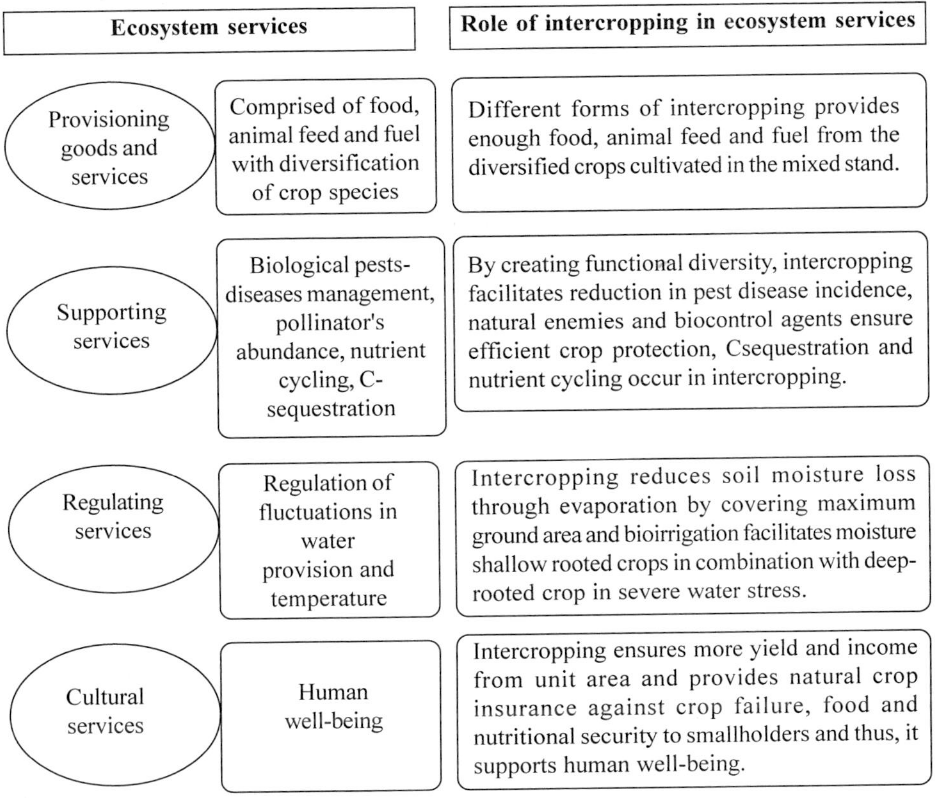

Figure 6.2. Role of intercropping in ecosystem services

6.3 Sustainable Agriculture

Sustainable agriculture focuses on the solutions for various problems associated with crop production. The main problems of present time in agriculture are dwindling and polluting natural resources (mainly land and water), loss of top soil by erosion and fertility, excessive use of chemicals, low productivity and poor farm output and inefficient marketing facility in the rural areas. The sustainable agriculture guides a clear way through some strategic approaches (Figure 6.3) as agriculture becomes dynamic and meaningful for the present and future generations.

Definition of sustainable agriculture

"Sustainable agricultural development is the management and conservation of the natural resource base, and the orientation of technological and institutional change in such a manner as to ensure the attainment and continued satisfaction of human needs for present and future generations. Such development... conserves land, water, plant and animal genetic resources, is environmentally nondegrading, technically appropriate, economically viable and socially acceptable."

Source: FAO, 2021a

5 Key principles for agricultural sustainability

Figure 6.3. FAO's definition of sustainable agriculture and five key principles of agricultural sustainability

Sustainable agriculture is such a system where proper care of natural resources are taken without any intention of over exploitation, rather it works with nature in an ecofriendly manner as the outcome is reflected into the qualitative improvement of agroecosystem. Moreover, FAO (2019) suggested that "sustainable agriculture must fulfill the requirements of present and future generations in terms of various products and services by assuring profitability, environmental health, and social and economic equity."

In principle, sustainable agriculture strives for using nature as the prototype for creation of different systems. Sustainable agriculture is designated mainly as farming system approaches and is capable of maintaining continuous farm productivity. Further, sustainable agriculture focuses on resource conservation, social support, soundness of environment and economic benefit. Sustainable agriculture can be assumed broadly as scientific ecosystem services (Altieri, 1995) which is more accomplished in the efficient use of natural resources, and applied inputs for enhancement of production as well as the benefit of people, and is in equilibrium with the agroecosystem without losing soil fertility. Presently, the approach to maximize agricultural production is synonymous to an enhancement in crop productivity per unit area and it needs the creation of efficient and sustainable cropping systems inclusive of intercropping as intercropping offers various qualities related to agricultural sustainability. Under properly managed conditions, intercrops can exhibit more yield and income than growing of sole crops. Moreover, it assures another social responsibility with more employment opportunities which is a key indicator of agricultural sustainability.

6.4 Intercropping and Agricultural Sustainability

The huge application of chemical inputs for nutrient management and plant protection is very common in industrialized agriculture and adoption modern agricultural technologies with monocropping system has received a global concern. Presently, it has been realized that monocropping is not the right way for attaining agricultural sustainability, because enough of chemical inputs are used in industrialized monoculture which causes pollution and creates imbalance in the biodiversity. In other words, it may be stated that intercropping as a type of diversified agricultural practice can be exploited to obtain agricultural sustainability as it is one of the possible ways to enrich multiplicity in a crop growing environment. The concept of agroecosystems is based on the understanding of functioning and mutually complementary connections between living organisms and their environment, which maintain a dynamic balance in time and space (Wiggering *et al.*, 2016). The cultivation of two or more crops simultaneously in mixed stands should be based on some ecological principles. Intercropping system assures ecological balance, enrichment of above- and below-ground diversity, higher deployment of available resources, maximization of productivity and ease of harm due to biotic agents (Yildirim and Eikinci, 2017; Maitra *et al.*, 2019). In an intercropping system, depending on their morphological and genetic structure, crops of different types are cultivated. However, the impact of allelopathy can be considered in the mixed stands as allelochemicals as exudated from one crop species can show negative influence on another; hence, crops selection is important in intercropping (Sharaby *et al.*, 2015; Hikal *et al.*, 2017).

Besides, agricultural sustainability directs us towards higher productivity for fulfillment of the needs of the population without ecological disturbance. Among different advantages of intercropping system, the most lucrative one is enhanced crop yields than pure stand. In many cases, due to complementarity between crops ensure maximum utilization of the limited resources and increase yield is obtained in the intercropping systems (Willey, 1979; Maitra *et al.*, 2019). Yield advantage is obtained in intercropping not only because of greater resource utilization, but also due to efficient use of applied nutrients and other agronomic management. (Bybee-Finley *et al.*, 2018). But there could also be expressions of competitive ability for growth resources among the crops cultivated in the mixture. Moreover, intercropping has been noted to reduce the risk of crop damage under any biotic or abiotic stress conditions and thus guarantees the stability of crop yield over time and assured income (Bybee-Finley *et al.*, 2016; Raseduzzaman and Jenson, 2017).

Intercropping is synonymous to improvement of soil health and cuts environmental pollution by reducing nitrogen losses (Sanderson *et al.*, 2013). Increased plant diversity maximizes beneficial soil microorganism population in the rhizosphere facilitating higher nutrient uptake by crops and reducing the population of harmful microbes living in the soil (Vukicevich *et al.*, 2016). Plants are not the standalone entities of agro-ecosystem as they host different micro-organisms in the rhizosphere and growing environment. All biotic agents (including endophytes and ectophytes) play their significant role in the crop environment. The host plants' biology, plant microbiome and their functioning as well as the catalyzing roles of micro-organisms different plant species are also important. The interaction between micro-organisms and plants may be symptomless (Han-ming *et al.*, 2019), but it may be pronounced in terms of enhancement of productivity and enrichment of soil microbial diversity. The intercropping system creates microbial diversity in the soil.

Although increased plant diversity is often related to efficient use of resources, considerable benefits may be added by intercropping even two crops, predominantly from a mix culture of grasses and legumes. The grasses claim more nitrogen and legumes do not require it. Legumes can fix atmospheric nitrogen and share the fixed nitrogen with non-legume crops in association and in this way, mixture of grass and legume in the intercropping system regulates the N-needs based on availability in soil. Such self-regulation is beneficial in terms of minimization of leaching of nitrate and de-nitrification, which are major providers to evils related to water quality and greenhouse gas discharges, respectively (Tang *et al.*, 2017).

Intercropping systems can regulate the biotic agents in crop fields. On-farm biodiversity enables a natural regulation against the population of harmful insects, pathogens and weeds. Therefore, less chemical is used for plant protection which further provides a congenial environment to beneficial organisms. In intercropping system, the population of beneficial insects such as predators and parasites is increased which can maintain the pest population dynamics. Hence, costs involved in plant protection and use of noxious chemicals are reduced which in turn check the pollution in the agroecosystem. Moreover, intercropping can create the host and habitat complexity of insects as two or more crops are grown simultaneously. The management of plant diseases is also simple in intercropping as crops are grown in mixture provide functional diversity that limits spreading out of pathogen due to difference in adaptation because of presence of different kinds of pathotypes. However, intercrops show advantages in weed management over sole crops because of more coverage of land area by crop combinations (Olorunmaiye, 2010) and / or defeat of the growth of unwanted weeds through the effect of allelopathy. In an intercropping system, more land area is occupied by the crops and so growth of weeds is suppressed.

There is enough scope for creation of diversity in agriculture in the form of crop rotation, cropping system, multiple cropping, mixed and intercropping. But intercropping offers a temporal and biological crop interaction (Han-ming *et al.*, 2019). Ultimately, harmony between the nature and crop management is the key consideration where plants can grow under favourable ecological makeup targeting sustainability. In this regard, an intercropping system can be well-thought-out as an appropriate option where a high degree of ecosystem service is maintained. In sustainable agriculture, nature and agroecosystem is nurtured without making a major change of landscape and avoiding much reliance on out-sourced inputs. Actually, sustainable agriculture creates usefulness of farming to society. An efficient resource utilization and socially supportive agriculture are the key factors for attaining sustainability. In intercropping system, available resources are fully utilized and it provides food and nutritional security to smallholders with employment opportunity which is the indicator of connecting society with farming. One should remember that sustainable agriculture not only assures continuous production with proper care of agroecosystem, but it provides commercially competitive farming under environmentally sound conditions in which above and below ground diversity can play a vital role.

6.5 Intercropping and Sustainable Development Goals

The UN in September 2015, adopted 17 Sustainable Development Goals (SDGs) which were ambitious agenda targets to end poverty and secure the world, for people, planet and prosperity and these goals are to be achieved by 2030. Intercropping has enough potential for poverty alleviation, reduction of hunger, provisioning of healthy foods and biodiversity enhancement covering directly or indirectly some SDGs, such as "**SDG 2** (end hunger, achieve food security and improved nutrition and promote sustainable agriculture), **SDG 13** (take urgent action to combat climate change and its impacts) and **SDG 15** (protect, restore and promote sustainable use of terrestrial ecosystems, sustainably manage forests, combat desertification, and halt and reverse land degradation and halt biodiversity loss)" (UN, 2022). Intercropping offers multifaceted advantages such as enhanced yield from unit area, higher use efficiency of resources (soil nutrients, soil moisture, atmospheric CO_2, sunlight and land), conservation of resources and soil health management and enhancement of soil fertility (by checking erosion and nutrient loss with run-off of water) and agricultural sustainability (diversification, legume effect, less chemical use and ecosystem services). The potential of intercropping in achieving the above SDGs is presented in the following figure (Figure 6.4).

SDGs	Salient features of SDG	Role of intercropping in achieving SDGs
	End hunger, achieve food security and improved nutrition and promote sustainable agriculture	More yield, production of nutrient rich and balanced food in intercropping cereals and legumes, promotion of agricultural sustainability are key benefits of intercropping.
	Take urgent action to combat climate change and its impacts	Intercropping is a nutrient efficient system and legumes in intercropping with other species reduces N-needs and thus, reduces GHGS emission in N-fertilizer production. Csequestration is beneficial to mitigate ill effects of climate change.
	Protect, restore and promote sustainable use of terrestrial ecosystems, sustainably manage forests, combat desertification, and halt and reverse land degradation and halt biodiversity loss	Intercropping facilitates greater ecosystem services, diversity in agriculture, soil conservation, reduced nutrients loss from top soil, and improvement of soil physico-chemical and biological properties.

Figure 6.4. Potential of intercropping system in achieving SDGs

Intercropping systems can be regarded as by interactions among the crops which will have an effect on growth and yield of plant species adopted and greater productivity. Better understanding of interactions among plant species and higher resource use potency in agro ecosystems are noted with intercropping. The impulses of sustainable practices in agriculture and also the ecological concerns obtaining from modern agricultural practices have had intercropping systems reassessed. The extent of accelerating crop productivity through appropriate intercropping systems has nonetheless been utilized to its full potential. Complementarities in resource use ought to be considered to induce achieving productivity in intercropping systems without any ill effects on agro-ecosystem. Sustainable agriculture seeks to use nature as the model for planning agriculture systems targeting uninterrupted production for the present and even for the future. Since nature consistently integrates & maintains diversity with the maximum level of ecosystem services, the intercropping system additionally creates the same by maintaining diversity in cropping & facilitates basic principles of sustainability.

References

Altieri, M. A. 1999. The ecological role of biodiversity in agro-ecosystems. *Agr. Ecosyst. Evviron.* **74**:19-31.

Altieri, M.A., 1995. Agroecology: the science of sustainable agriculture, second edition. Publisher: Westview Press.

Bybee-Finley, K.A. and Ryan, M.R., 2018. Advancing intercropping research and practices in industrialized agricultural landscapes. *Agric.* **8**:80.

Bybee-Finley, K.A.; Mirsky, S.B. and Ryan, M.R. 2016. Functional diversity in summer annual grass and legume intercrops in the northeastern United States. *Crop Sci.* **56**:2775.

Emerson N. 2007. Cropping systems; Illinois Agronomy Handbook. Pp 49-50. ednaf@illinois.edu.

FAO. 2019. Sustainable Development Goals, Overview: FAO and the Post-2015 Agenda, Sustainable Agriculture, www. fao.org (last visited: 15 June, 2019).

FAO. 2021a. Sustainable agriculture and rural development. https://www.fao.org/3/u8480e/U8480E0l.htm; (accessed 25 December, 2021).

FAO. 2021b. Sustainable food and agriculture. https://www.fao.org/sustainability/en/; (accessed 25 December, 2021).

Han-ming, H., Li-na, L., Munir, S., Bashir, N. H., Yi, W., Jing, Y. and Cheng-you, L. 2019. Crop diversity and pest management in sustainable agriculture. *J. Int. Agric.* **18**(9):1-1952.

Hauggaard-Nielsen, H., Ambus, P. and Jensen, E. S. 2001. Interspecific competition, N use and interference with weeds in pea-barley intercropping. *Field Crops Res.* **70**:101-109.

Hikal W. M., R.S. Baeshen, H.A.H. Said-Al Ahl. 2017. Botanical insecticide as simple extractives for pest control. *Cogent Biol.* **3**:1-14.

Hole, D. G., Perkins, A. J., Wilson, J. D., Alexander, I. H., Grice, P. V., Evans, A. D. 2005. Does organic farming benefit biodiversity? *Biol. Conserv.* **122**:113–130.

Jackson, L.E., Pascual, U. and Hodgkin, T. 2007. Utilizing and conserving agro-biodiversity in agricultural landscapes. *Agr. Ecosyst. Environ.* **121**:196-210.

Lal, R. 2004. Agricultural activities and the global carbon cycle. *Nutr. Cycl. Agroecosyst.* **70**(2):103–116.

Li, L., Sun, J. H., Zhang, F.S., Li, X.L., Yang, S.C., et al. 2001. Wheat/maize or wheat/soybean strip intercropping: I. Yield advantage and interspecific interactions on nutrients. *Field Crops Res.* **71**:123–137.

Lichtfouse, E., Navarrete, M., Debaeke, P., Souchère, V., Alberola, C. and Ménassieu, J. 2009. Agronomy for sustainable agriculture, A review. *Agron. Sustain. Dev.* **29**:1-6.

Maitra, S. and Ray, D. P. 2019. Enrichment of biodiversity, influence in microbial population dynamics of soil and nutrient utilization in cereal-legume intercropping systems: A review. *Int. J. Biores. Sci.* **6**:11–19, doi:10.30954/2347- 9655.01.2019.3

Maitra, S., Hossain, A., Brestic, M., Skalicky, M., Ondrisik, P., Gitari, H., Brahmachari, K., Shankar, T., Bhadra, P., Palai, J.B., Jena, J., Bhattacharya, U., Duvvada, S.K., Lalichetti, S., Sairam, M. 2021. Intercropping- a low input agricultural strategy for food and environmental security. *Agron.* **11**:343; https://doi.org/10.3390/ agronomy11020343.

Maitra, S., Palai, J.B., Manasa, P., Prasanna Kumar, D. 2019. Potential of intercropping system in sustaining crop productivity. *Int. J. Agric. Environ. Biotechnol.* **12**(1):39-45.

Maitra, S., Shankar, T. and Banerjee, P. 2020. Potential and advantages of maize-legume intercropping system (Online First), Intech Open, DOI: 10.5772/intechopen.91722. Available from: https://www.intechopen.com (Accessed on 18 March, 2021).

McMichael, A. J., Powles, J. W., Butler, C. D., and Uauy, R. 2007. Food, livestock production, energy, climate change, and health. *Lancet.* **370:**1253–63.

Olorunmaiye, P.M. 2010. Weed control potential of five legume cover crops in maize/cassava intercrop in a Southern Guinea savanna ecosystem of Nigeria. *Aust. J. Crop Sci.* **4**:324-329.

Raseduzzaman, M. and Jensen, E.S. 2017. Does intercropping enhance yield stability in arable crop production? A meta-analysis. *Eur. J. Agron.* **91**:25–33.

Sanderson, M.A., Archer, D.; Hendrickson, J.; Kronberg, S.; Liebig, M.; Nichols, K.; Schmer, M., Tanaka, D.; Aguilar, J. 2013. Diversification and ecosystem services for conservation agriculture: Outcomes from pastures and integrated crop–livestock systems. *Renew. Agric. Food Syst.* **28**:129–144.

Scherr, S. J. and McNeely, J. A. 2008. Biodiversity conservation and agricultural sustainability: towards a new paradigm of 'ecoagriculture' landscapes. *Philos. Trans. Royal. Soc. B.* **363**:477-494.

Sharaby A., H. Abdel-Rahman, S. Sabry Moawad. 2015. Intercropping system for protection the potato plant from insect infestation. *Ecologia Balkanica*. **7**(1):87-92.

Tang, Y., Yu, L., Guan, A., Zhou, X., Wang, Z., Gou, Y. and Wang, J. 2017. Soil mineral nitrogen and yield-scaled soil N_2O emissions lowered by reducing nitrogen application and intercropping with soybean for sweet maize production in southern China. *J. Integr. Agric.* **16**:2586–2596.

Tilman, D., Cassman, K. G., Matson, P. A., Naylor, R. and Polasky, S. 2002. Agricultural sustainability and intensive production practices. *Nature*. **418**:671-677.

UN. 2022. The 17 Goals, Department of Economic and Social Affairs, Sustainable Development. https://sdgs.un.org/goals (Accessed 25 December, 2021).

Vukicevich, E., Lowery, T., Bowen, P., Urbez-Torres, J.R. and Hart, M. 2016. Cover crops to increase soil microbial diversity and mitigate decline in perennial agriculture. A review. *Agron. Sustain. Dev.* **36**:48.

Wiggering H., Weißhuhn, P. and Burkhard, B. 2016. Agrosystem Services: An Additional Terminology to Better Understand Ecosystem Services Delivered by Agriculture. *Landscape Online*. **49**:1-15.

Willey, R.W., 1979. Intercropping- its importance and research needs. Part 1: Competition and yield advantages. *Field Crops Abst.* **32**:1-10.

Willey, R.W., Natarajan, M., Reddy, MS., Rao, M.R., Nambiar, P.T.C., Kannaiyan, J. and Bhatnagar, V.S. 1983. Intercropping studies with annual crops. Pages 83-100, In: Bette crops for food. Pitman, London, U.K.

Yildirim, E. and Eikinci, M. 2017. Intercropping Systems in Sustainable Agriculture. *Süleyman Demirel Üniversitesi Ziraat Fakültesi Dergisi.* **12**(1):100-110.

7

Alley Cropping

In the current situation, crop intensification is of topmost importance to increase agricultural production from ever-limiting and degrading resources. In this regard, the potential of agroforestry can be thought that offers ample opportunity to enhance farm output with multifaceted benefits (Xu *et al*., 2019). Forests are known to maintain ecological balance in many ways with greater ecosystem services; hence, the advantages of forestry can be included in agroecosystem in the form of agroforestry which enables yield of forest products and crops together. Among various agroforestry systems, alley cropping (Figure 7.1) is an option where crops can be grown harnessing the benefits of forests (Solaimalai *et al*., 2005). In the sloppy and erosion prone, low fertile and marginal lands, successful raising of crops is a difficult task, but alley cropping, that is, the combination of selected crops and forests can ensure economically viable farm productivity by offering short-, medium- and long-term needs. As it has been mentioned earlier that more farm output is required to meet the demand for the future, fertile and productive lands are already engaged. There is further scope to bring less-fertile and marginal lands into crop production by adoption of agroforestry and thus, alley cropping can be adopted.

Figure 7.1. Alley cropping of perennial tree and annual crop

Alley cropping is a form of intercropping and also termed as hedgerow intercropping where food, forage and specialty crops are cultivated in the space available between rows of perennial crops. Similar to other intercropping systems, two or more different types of crops coexist in alley cropping and appropriate agronomic management is to be provided to all component crops. In general, annual food crops or forages are grown in the alleys (free space) between two rows of perennial plants (hedgerows) which may be trees or shrubs (Korwar, 1990). Crop choice and other management aspects are important for the success of an alley cropping.

7.1 Management of Crops in Alley Cropping

Tree species are cut frequently and these should be kept pruned throughout the growing period of annual crops to minimize shading and to reduce competition with food crops; however, they are allowed to grow freely to cover the land, when there is no food crop. The alley management mainly includes choice of crops (tree, shrubs and annual crops), maintenance of alley width, pruning and nutrient management.

7.1.1 Choice of Crops

The crops considered in alley cropping are tree, shrubs and annuals. Tree and shrubs are perennials planted in alleys and annuals (generally food or forage crops) are grown in between two alleys. The woody tree and shrub species commonly chosen for alley cropping are: *Acacia auriculiformis*, *Acacia barterii*, *Albizia lebbeck* (saras or frywood), *Alchomea cordifolia*, *Azadirachta indica* (neem), *Cajanus cajan*, *Desmanthus virgatus*, *Dalbergia sissoo* (sisu or north Indian rose wood), *Dactyladenia barteri*(syn. *Acioa barteri*), *Calliandra calothyrsus*, *Erythrina poeppigiana*, *Flemingia congesta*, *Flemingia macrophylla*, *Gliricidia sepium*, *Gmelina arborea*, *Leucaena leucocephala* (subabul), *Paraserianthes falcataria* (syn. *Albizia falcataria*) etc. In tropical north and south America, *Inga edulis* and *Inga oerstediana* are selected for hedgerows. On basic soils, *Leucaena leucocephala* and *Glificidia sepium* perform well, but on acidic Ultisols and Oxisols, growth of these plants is reduced because of low phosphorus and calcium and more of aluminum. *Acacia angustissima* and *Calliandra calothyrsus* are chosen in the western highlands of Cameroon (Akume *et al*., 2015).

The common annual crops grown in alley cropping are cereals (maize, sorghum, wheat, upland rice), leguminous crops (beans, cowpea, soybean), root crops (cassava, yam) and different types of vegetables. Companion crops are sown in the alleys between the hedgerows. The choice of companion crop is highly region specific and it depends greatly on the choice of the farmers. Generally, food and forage crops are grown in alleys. Moreover, fruits and specialty crops are also considered and sometimes biomass producing crops are also chosen. During the initial period of hedgerow establishment, there will be less shades and food crops can be easily grown. But gradually trees and shrubs will grow. Though these will be pruned periodically, these will influence the growing environment and alley crops may get limited resources. At that time growing of shade tolerable crops and specialty crops will be wise. Integration of specialty crops is also a practice in alley cropping. The common specialty crops are fruits and vegetables, tree nuts, horticulture and nursery crops. When the hedgerows will be fully grown and after pruning also providing shade, it is better to grow shade loving or tolerant specialty crops in alleys.

7.1.2 Crops Establishment

In sloppy lands, the perennial crops are grown by direct seeding or transplanting of seedling along the contour or across the slope. In arable areas when wind erosion control is one of the objectives of alley cropping, it is better to adopt closer plant to plant spacing within the row for tree species. Sometimes closer intra hedgerow planting is done in trenches rather than in holes or pits. Later on the basis of requirement close trees may be thinned. After establishment of tree species, regular agronomic management is essential for proper growth and desired yield from hedgerows. The trees species for hedgerows should be of faster growth, less mortality, faster regrowth after repeated purnings, ability to fix N biologically (legumes), tolerant to abiotic stress and suitable for multipurpose use (Solaimalai *et al.*, 2005). Further, the tree species should have fewer roots at the soil surface to reduce the competition with crops grown in the alleyway (Hodge *et al.*, 1999).

In case of annual food crops, direct sowing is commonly observed and dense seeding is advised. At the early seedling stage, extra plants are thinned. But for direct seeding, a good seed bed is required to facilitate germination and uniform stand. In the humid and high rainfall areas, direct seeding is preferred, however, in dry zones, transplanting of seedling is advised for proper stand establishment. The crops should be of short duration, ecologically hardy, wider adaptability to different climatic conditions, tolerant to partial shade and abiotic stress and shallow rooted.

7.1.3 Alley Width

In alley cropping, if the alley width is narrow, there is a possibility of maximum competition between hedgerow crops and annual crops grown. The tree and/or shrub rows are planted on the basis of the purpose. The planting geometry depends on different factors like growth potential of hedgerows, whether single or multiple species to be planted, whether single or multiple rows to be planted, length of the slope, breadth of the field, requirement of light for the crops grown in between alleys and farmer's choice. In semi-arid regions where moderately good amount of rainfall is received, wider alleys are recommended and considerably good return is obtained from trees as well as annual crops (Singh et al., 1989). Hedgerows of *Leucaena leucocephala* yielded less with wide spacing and more yield was with narrow alley width (4.0 m) over wider alley width of 6.0 m as mentioned by (Solaimalai *et al.*, 2005). When food production is the major goal for smallholders, wider hedgerow width is preferred to accommodate more annual food crops in between alleys. The total production of biomass will be more in closed spacing of alleys. In areas with low rainfall, wider alleys are planted to avoid completion for soil moisture. Wider alleys reduce shading to annual agricultural crops. In a study, the wheat yield remained on par in both of *Gliricidia sepium* (3.63 t ha^{-1}) and *Leucaena leucocephala* (3.31 t ha^{-1}) planted in 6 m alley width in Bangladesh (Ferdush *et al.*, 2018). Madhu *et al.* (2019) suggested for planting of tree crops *Gliricidia maculata* and *Leucaena leucocephala* with 5-10 m wide is suggested in eastern region of India. In case of *Azadirachta indica* (neem) 8 m spacing between two rows can be given to minimize competition with annual crops. In temperate areas, there is limited scope for adoption of alley cropping and in general a food crop is grown with timber species (Wolz and DeLucia, 2018). However, Wolz *et al.* (2018) suggested diversification of woody species, and alley cropping can be adopted with components like food crops and fodders in temperate regions.

7.1.4 Pruning Height and Frequency of Pruning

Pruning height is an important aspect of management of hedgerows. When the annual crops will be cultivated, it is essential to prune the tree crops to minimize the shading effect. A low pruning height of 15-30 cm to *Leucaena* would be desirable to evade shading the annual crop with low canopy height. A low pruning height of 20-40 cm for *Leucaena* caused its wood to split and make it prone to termite attack. In a study, it was noted that the trees survived with as short as 7.5 cm stumps with greater herbage yield than taller stumps (Korwar, 1990). In general, pruning height of 30-90 cm is satisfactory. The biomass production of *Leucaena leucocephala* was more when pruned at a height of 90 cm as compared to 60 cm and 30 cm (Maiti and Chattarjee, 1986). Actually, the pruning height depends on the basis of adoption of component crops in the mixed stand. The lower the hedgerow and taller the crop, the less frequently pruning is needed. The main objective of pruning is to facilitate sunlight to both the crops and maximum use of it.

The time of the beginning of pruning the hedgerows is important in low fertile lands. In Hawaii, *Leucaena* harvested when four months old crop gave the maximum leaf yield (at height of 120-150 cm) and there was a gradual decrease in yield with increase of plant height and cutting delay (Korwar, 1990). But in Philippines, the first cutting of *Leucaena* is preferred when plants are six months old. In semi-arid regions *Leucaena* needs more time to give a sizable yield and generally takes about a year. The fast-growing crops like *Gliricidia* and *Leucaena* require pruning at every five to six weeks during cropping season. Further, too low or two frequent pruning should be avoided as there will be a chance of dieback disease infection.

Full grown canopy of *Leucaena* alley during lean period

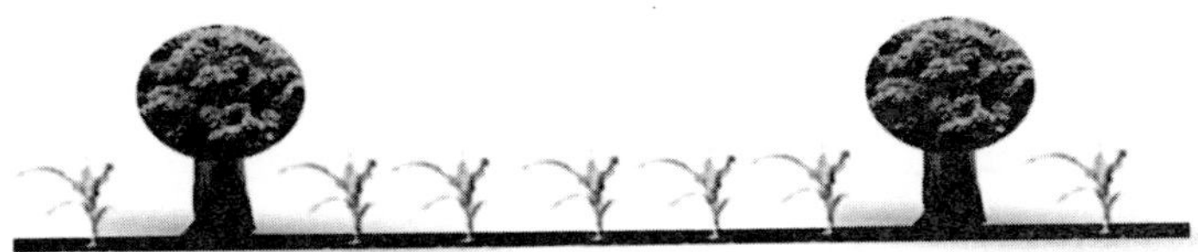

Prunned *Leucaena* alley two weeks after maize sowing

Alley cropping of maize at full grown stage in *Leucaena*

Figure 7.2. Schematic drawing showing management of *Leucaena* and maize alley cropping

7.1.5 Root Pruning

Interaction between tree species and annual crops is very important to get benefit from the mixed cropping. There will always be competition for available resources even during the early stage of establishment of tree species and the competition may decrease yield of annual crops because both the species will absorb soil moisture and nutrients from the common pool (Noordwijk *et al.*, 2015). To reduce the competition, root pruning of tree crops is practiced in alley cropping. Root pruning is done by tilling the soil with country plough up to 30 cm depth on both sides of hedgerows. In India, root pruning is done during monsoon season. To restrict the horizontal growth of roots, a polythene sheet as root barrier may be placed at 0.3 m apart the hedgerow trunk and up to 0.5 to 1.2 m depth (Singh *et al.*, 1989; Gillespie *et al.*, 2000; Solaimalai *et al.*, 2005). In case root pruning is not adopted, the tree species will spread roots in between the gap of alleys and yield of annual crop will be reduced.

7.1.6 Nutrient Balance and Management

The legume hedgerows contribute more nutrients (especially N) than non-legumes and annual crops are benefited. On an average the biomass production of *Leucaena* varies from 1.1 to 3.5 t ha^{-1} $year^{-1}$ and the quantity of N and P recycled will be 33-150 and 2.2 to 7.0 kg respectively. The entire quantity of recycled nutrients are not used by annual crops and it is noted that annual crops get only 10 to 20 % of N in a given season (Solaimalai *et al.*, 2005). Nutrient yield of some common tree species is tabulated as follows (Table 7.1).

Table 7.1 Nutrient yield from five prunings of hedgerows at International Institute of Tropical Agriculture, Nigeria (4m x 0.5 m spacing)

Species	Nutrient yield kg ha^{-1} $year^{-1}$				
	N	P	K	Ca	Mg
Alchornea cordifolia	85	6	48	42	8
Dactyladenia (Acioa) barteri	41	4	20	14	5
Gliricidia sepium	169	11	149	66	17
Leucaena leucocephala	247	19	185	98	16

Source: Nair, 1993

In Kenya, *Leucaena* prunings added 52 kg N and 3 kg P ha^{-1} $season^{-1}$ to the soil, but maize + *Leucaena* alley cropping caused a positive balance of 15 kg for N and a negative balance of 4 kg for P ha^{-1} $season^{-1}$. Under this situation, negative P balance could be corrected by application of cattle manures (Lupwayi and Haque, 1999). In dry tropical Savannah of Northern Kenya, a low nutrient input region, it was noted enhancement N and K balance in alley cropping system of *Acacia*-sorghum and a nutrient return was observed due to mulching crop residues and *Acacia* leaves which resulted in a positive nutrient balance (Lehmann *et al.*, 1999). In Malawi, *Gliricidia sepium* + maize alley cropping system, maize yielded 3.3 t ha^{-1} as compared to one 1.0 t ha^{-1} in pure stand of maize without application of any chemical nutrients (Akinnifesi *et al*., 2006) which indicated the benefits of tree species in providing nutrients to annual crops. Several studies clearly indicated that pruning and leaf fall can add moderately good quantities of N, but for the requirement of other nutrients, there is need for exogenous application to achieve a good harvest from annual crops in alley cropping. Beule *et al.* (2019) observed that in Germany, cereals and oilseeds raised in 12 m wide poplar hedgerows in temperate conditions also altered the fullness of soil micro-organisms and soil-N-cycling genes facilitating soil fertility.

7.2 Benefits of Alley Cropping

Alley cropping provides an ideal diversification with the fullest scope for utilization of available resources and the increase of short, medium and long term returns inclusive of regular cash flow from marginal lands. Like other intercropping systems, it offers other benefits which are discussed below.

7.2.1 Greater use of Resources

Like other intercropping systems in alley cropping also, the basic principles are followed as two or more crops are cultivated together. The maximum utilization of space (land area) is ensured as more plants are put-up in a unit area and other natural resources such as soil moisture, atmospheric carbon dioxide and sunlight are better utilized than pure stands of forestry. The poor and marginal areas are chosen for alley cropping in general where commercial production of annual crops are not preferred. In this way, it adds value in marginal lands.

7.2.2 Performance of Crops and Yield Advantage

Earlier researches carried out worldwide evidenced that alley cropping was able to enhance crop productivity in most of the cases from the marginal lands. But yield advantage of crops depends on several factors, namely, growing situation of crops, impact of agroclimatic conditions, choice of crops and agronomic management provided. For example, in an experiment conducted at Jhansi, yield of sesame was increased in *Leucaena* alleys, however, peanut and red gram did not perform well (Singh, 1983) and it might be due to combination of legume (*Leucaena*) and nonlegume (sesame) in the specific condition. In contrast, perennial red gram when alley cropped with peanut, yield benefit was recorded compared to perennial red gram + sorghum alley cropping (Solaimalai *et al*., 2005). Alley cropping of cowpea yields more than sole cropping when grown in alleys of *Gliricidia* and *Leucaena* in Guyana (Chesney *et al*., 2010).

7.2.4 Regular Cash Flow

In degraded, poorly fertile and marginal lands, forestry can be adopted. But there is scope for adoption of different agroforestry systems inclusive of alley cropping. Alley cropping has the potential to register considerable farm output even from marginal and degraded lands. A combination of annual and specialty crops can be grown in alleys consisted of shrubs and perennial trees. In such case, specialty crops as well as annual crops will ensure regular cash flow (which is much required for small farmers) and other perennial crops will yield fodder, fuel and timber in medium- and long-term. The degraded and marginal lands are mostly managed by poor farmers where investment in agriculture is less and the regular cash flow helps the smallholders to operate farm efficiently and manage their livelihood.

7.2.5 Soil Conservation and Prevention of Nutrient Loss

The marginal and sloppy lands are highly prone to soil erosion erosivity (more than 8) even when annual crops are cultivated. In alley cropping, hedgerows act as barriers in restricting soil erosion. Further, the mulching is done with pruned leaves and twigs that also check soil erosion. In the sloppy lands (8-10% slope) of eastern India, planting perennial *Gliricidia maculata* and *Leucaena leucocephala* with 5 to 10 m alley width is suggested to prevent soil loss. The space between alleys is used for growing annual food crops or forages. Experimental results evidenced that alley cropping with *Gliricidia* declined runoff by 27.9-28.2% and soil loss by 49.3-51.1% over no alley cropping and *Leucaena* alley with miniature trench reduced runoff by 18.3-18.7% and soil loss by 37.2-43.0% (Madhu *et al.*, 2019). Moreover, alley cropping has enough potential to reduce nutrient loss from the top soil. Both of *Gliricidia* and *Leucaena* alleys are known to conserve nutrient loss by checking loss of topsoil. *Gliricidia* alleys conserved organic carbon, nitrogen, phosphorus and potassium by 63.4, 5.0, 0.3 and 2.4 kg ha^{-1}, respectively over no alley cropping. Similarly, *Leucaena* alleys were also equally important and, in an experiment, it conserved organic carbon, nitrogen, phosphorus and potassium by 57.7, 4.6, 0.3 and 2.2 kg ha^{-1}, respectively over no alley cropping (Madhu *et al.*, 2019).

7.2.6 Carbon Sequestration

Carbon sequestration of great importance in the present context of global warming and carbon footprint in agriculture. Alley cropping potentially registers caron sequestration and adding woody biomass into the soil and thus, it can be considered as an ecofriendly cropping system. The research findings evidenced that agroforestry increased carbon stock in dry regions of Kenya (Reppin *et al.*,

2019). In the Philippines, jatropha and corn alley cropping remarkably increased organic carbon stock in soil (Marin, 2016). Alley cropping with *Gliricidia maculata* and *Leucaena leucocephala* stored 4.26 and 4.52% more soil organic carbon, respectively over no alley cropping (Madhu *et al.*, 2019). In the temperate Mediterranean region also, alley cropping registered more carbon sequestration (Cardinael *et al.*, 2015).

7.2.7 Enrichment of Biodiversity

The biodiversity enrichment in alley cropping is observed in many ways (Xu *et al.*, 2019; Staton *et al.*, 2022). First, more and one plant species are grown together that enriches crops diversity than pure stand of crops. Secondly, alley cropping offers shelter and supports habitat of birds and bats (Harvey *et al.*, 2007). Thirdly, below-ground biodiversity is noted in the form of microorganisms, arthropods and other fauna. Ashraf *et al.* (2018) recorded the abundance of arthropods and species diversity into the soil where alley cropping was adopted.

7.2.8 Climate Change Mitigation and Adaptation

Alley cropping has a huge potential to sequester carbon as the tree spreads are regarded as to record more than three-fourth of global carbon pool (Zomer *et al.,* 2016). Reppin *et al.* (2019) reported that in Kenya, adoption of agroforestry increased carbon storage in soil and upgraded livelihoods of smallholder farmers. A sizable amount of carbon dioxide emission in agriculture is caused by anthropogenic interventions and carbon sequestration is in top priority for sustainable agriculture. Alley cropping sequester carbon in two ways, namely, inclusion of aboveground biomass (leaves, twigs etc.) and belowground biomass (roots). Thus, alley cropping as well as agroforestry is potentially considered as a climate-smart cropping system (Yirefu *et al.*, 2019). Presently, it is considered as one of the important adaptation strategies to minimize the adverse effects of climate change. Alley cropping can play a buffering role by creating microclimate under climatic extremes for yield stability and thus, it can be considered as an adaptation strategy against climate change (Reyes *et al.*, 2021).

7.3 Alley Cropping: Limitations and Remarks

Though there are several benefits, alley cropping also shows few limits (Figure 7.3) as multiple crops are grown at a time on the same land. If the limitations are critically and properly analyzed, possible solutions can be chalked out.

In resource-poor marginal lands investment in agriculture is less and farmers do not use costly inputs. Under this situation alley cropping may be considered as a suitable option of efficient land use and sustainable agricultural production. In some cases, alley cropping may yield less from annual crop than pure stand, but multifaceted benefits should be considered while evaluating. Proper planting geometry, choice of suitable crops and appropriate management can make alley cropping more profitable than sole crops as evidenced in different countries of the world.

Limitations	Remarks
More intensive management is needed including dedicated tools and equipment for the hedgerows and skills are required for managing multiple crops.	Once alley cropping is adopted, farmers should get well-acquainted with the skills and the required equipment needs to be arranged.
At the beginning, alley cropping may not result in lucrative return (as income comes from only annual crops).	Alley cropping is adopted for long term and from the marginal and problematic lands lucrative return is not always expected.
Alley cropping requires proper marketing of yield output from perennial crops.	Necessary marketing facilities are to be created and crops are to be chosen based on local demand
Tree species may create complications to annual crops.	Crop choice can solve the limitation in terms of greater compatibility. Crop choice can reduce the issue.
Interspecies competition may create problem.	Integrated weed management should be adopted.
Scope for herbicides application is limited which may increase cost of cultivation.	Further, it offers employment opportunities to underutilized family labourers for smallholders.

References

Akinnifesi, F. K., Makumba, W. and Kwesiga, F. R. 2006. Sustainable maize production using gliricidia/ maize intercropping in southern Malawi. *Exper. Agric.* **42** (4):10 doi:10.1017/ S0014479706003814.

Akume, N. D., Christopher, S., Manga, M. A., Francis, N. A. S., Yaya, F. V. and Venasius, L. 2015. Effect of tree hedgerow pruning on maize yield in Santa, Cameroon. *Int. J. Agric. Innov. Res.* **3**(6):1750-56.

Ashraf, M., Zulkifli, R., Sanusi, R., Tohiran, K.A., Terhem, R., Moslim, R., Norhisham, A.R., Ashton-Butt, A. and Azhar, B. 2018. Alley-cropping system can boost arthropod biodiversity and ecosystem functions in oil palm plantations. *Agric. Ecosyst. Environ.* **260**:19-26.

Beule, L., Corre, M. D., Schmidt, M., Go¨bel, L., Veldkamp, E. and Karlovsky, P. 2019. Conversion of monoculture cropland and open grassland to agroforestry alters the abundance of soil bacteria, fungi and soil-N-cycling genes. *PLoS ONE.* **14**(6):e0218779. https://doi.org/ 10.1371/journal. pone.0218779.

Cardinael, R., Chevallier, T., Barthès, B.G., Saby, N.P., Parent, T., Dupraz, C., Bernoux, M. and Chenu, C., 2015. Impact of alley cropping agroforestry on stocks, forms and spatial distribution of soil organic carbon-A case study in a Mediterranean context. *Geoder.* **259**: 288-299.

Chesney, P. E. K., Simpson, L. A., Cumberbatch, R. N., Homenauth, O. and Benjamin, F. 2010. Cowpea yield performance in an alley cropping practice on an acid infertile soil at Ebini, Guyana. *Open Agric. J.* **4**:80-84.

Ferdush, J., Karim, M. M., Himel, R. M., Saha, S. R. and Ahamed, T. 2018. Impact of Alley Cropping on Wheat Productivity. *IOSR J. Agric. Vet. Sci.* **11**(2):17-25.

Gillespie, A. R., Jose, S., Mengel, D. B., Hoover, W. L., Pope, P. E., Seifert, J. R., Biehle, D. J., Stall, T. and Benjamin, T. J. 2000. Defining competition vectors in a temperate alley cropping system in the mid-western USA. *Agroforestry Syst.* **48:**25-40.

Harvey, Celia A., Villalobos and Jorge A. González. 2007. Agroforestry systems conserve species-rich but modified assemblages of tropical birds and bats. *Biodiversity Conserv.* **16** (8):2257–2292. DOI:10.1007/s10531-007-9194-2.

Hodge, S., Garrett, H. E. and Bratton, J. 1999. Alley Cropping: An Agroforestry Practice, Agroforestry Notes (USDA-NAC). 11. available at: https://digitalcommons.unl.edu/agroforestnotes (accessed on: 24 April, 2020).

Korwar, G. R. 1990. Alley cropping, *In:* Intercropping principles and practices, (*Eds.* Hosmani, M. M., Chittapur, B. M. and Hiremath, S. M.), Proc. Of summer short course, Department of Agronomy, College of Agriculture, Dharwad, India, May 07-16, 1990, pp.155-160.

Lehmann, J., Weigl, D., Droppelmann, K., Huwe, B. and Zech, W. 1999. Nutrient cycling in an agroforestry system with runoff irrigation in Northern Kenya. *Agroforestry Syst.* **43**:49-70.

Lupwayi, N. Z. and Haque, I. 1999. Leucaena hedgerow intercropping and cattle manure application in the Ethiopian highlands. III. Nutrient balances. *Biol. Fert.Soils.* **28**:204-2 11.

Madhu, M., Hombegowda, H. C., Beer, K., Adhikary, P. P., Jakhar, P., Sahoo, D. C., Dash, C. J., Kumar, G. and Naik, G. B. 2019. Status of natural resources and resource conservation technologies in eastern region of India, (In: Karma Beer *et al., Eds.*) *Resource conservation in eastern region of India: lead papers of FFCSWR*, 2019, Indian Association of Soil and Water Conservationists, Dehradun, Uttarakhand, pp.60-82.

Maiti. S. and Chatterjee. B.N. 1986. In: Proc. of National Seminar on Recent Trends in Forage Production. IGFRI. Jhansi, India, pp. 88-89.

Marin, R.A. 2016. Jatropha-based alley cropping system's contribution to carbon sequestration. *Int. J. Agron. Agric. Res.* **8**(1):1-9.

Nair, P. K. R. 1993. An introduction to agroforestry. Kluwer Academic Publishers, the Netherlands, ISBN: 0-7923-2134-0. p. 489.

Noordwijk, M. van, Lawson, G., Hairiah K and Wilson, J. 2015. Root distribution of trees and crops: competition and/ or complementarity. *In*: Ong, C.K., Black, C.R., Wilson, J., *Eds.* Tree-crop interactions: agroforestry in a changing climate. 2nd Ed. Wallingford: CABI; pp. 221–257. https://doi.org/10.1079/ 9781780645117.0221.

Reppin, S., Kuyah, S., de Neergaard, A., Oelofse, M., Rosenstock, T. S. 2019. Contribution of a groforestry to climate change mitigation and livelihoods in Western Kenya, *Agroforest Syst.* **94**:203-220. DOI:10.1007/s10457-019-00383-7.

Reyes, F.; Gosme, M.; Wolz, K.J.; Lecomte, I.; Dupraz, C. Alley cropping mitigates the impacts of climate change on a wheat crop in a Mediterranean Environment: A biophysical model-based assessment. *Agric* 2021, **11**: 356. https:// doi.org/10.3390/agriculture11040356.

Singh, R. P., Van den Beldt, R. J., Hocking, D. and Korwar, G.R. 1989. Alley farming in the semi-arid regions of India, In: Proc. Int. Workshop on Alley Farming in the Humid and Sub-humid Tropics (*Eds.* B.T. Kang, B.T. and Reynolds, L.), March 10-14, 1989, Nigeria, pp. 108-122.

Singh, G. B. 1983. Role of Agroforestry in improving the environment. *Indian Fmg.* **33**: 15-19.

Singh, R. P., Ong, C.K. and Saharan, N. *1989.* Above- and below-ground interactions in alley cropping in semiarid India. *Agroforest Syst.* **9**: 259–274.

Solaimalai, A., Muralidaran, C. and Subburamu, K. 2005. Alley cropping in rainfed agro-ecosystem - A review. *Agric Rev.* **26**(3):157-172.

Staton, T., Breeze, T. D., Walters, R. J., Smith, J. and Girling, R. D. 2022. Productivity, biodiversity trade-offs, and farm income in an agroforestry versus an arable system. *Ecol. Econ.*, **191**:107214. https://doi.org/10.1016/j.ecolecon.2021.107214.

Wolz, K. J. and DeLucia, E. H. 2018. Alley cropping: Global patterns of species composition and function. *Agric. Ecosyst. Environ.* **252**:61–68. DOI:10.1016/j.agee.2017.10.005.

Wolz, K. J., Lovell, S. T., Branham, B. E., Eddy, W. C., Keeley, K., Revord, R. S., Wander, M. M., Yang, W. H. and DeLucia, E. H. 2018. Frontiers in alley cropping: Transformative solutions for temperate agriculture. *Global Change Biol.* **24:**883–894.

Xu, H., Bi, H., Gao, H. and Yun, L. 2019. Alley cropping increases land use efficiency and economic profitability across the combination cultivation period, *Agron.* **9:**34. DOI:10.3390/agronomy9010034.

Yirefu T, Yenenesh H, Zemedkun S. 2019. Potential of agroforestry for climate change mitigation through carbon sequestration: Review paper. *Agri. Res. Tech. Open Access J.* **22**(3): 556196. DOI: 10.19080/ARTOAJ.2019.22.556196.

Zomer, R. J., Neufeldt, H., Xu, J., Ahrends, A., Bossio, D., Trabucco, A., Van Noordwijk, M., Wang, M. 2016. Global tree cover and biomass carbon on agricultural land: The contribution of agroforestry to global and national carbon budgets. *Sci. Reports*. **6** (1): 29987. DOI:10.1038/srep29987.

8

Legumes in Intercropping and Soil Improvement

Enhancement of population is a global issue and the expected world population will be 9.7 billion in 2050 (UN, 2019). Under the situation, achievement of food and nutritional security is a major target with reduction of GHGs to the atmosphere and mitigation of ill effects of climate change. Modern agriculture is responsible for emission of GHGs in many ways and one of them is production of synthetic N fertilizer. There is no doubt about the fact that chemical N fertilizers enhance crop yield but it is at the cost of atmospheric equilibrium and agricultural sustainability cannot be achieved by this practice (Bedoussac *et al.* 2015; Layek *et al.*, 2018). There is urgent need for reduction of emission of GHGs and creation of biodiversity for achieving agricultural sustainability (Maitra and Ray, 2019). Intercropping or polyculture is one of the options for creation of biodiversity in crop field and use of chemical N can be minimized by inclusion of legumes in intercropping system. Cereals are energy crops and protein-rich legumes provide nutrition. The combination of cereal + legume is the most widely practiced intercropping system.

Legumes are the plants (family: *Fabaceae* or *Leguminosae*) cultivated mainly as food crops, forage for feeding of livestock and making of silage and green manure crops. Cultivation of legume as a component in intercropping system has multi-dimensional benefits as the crops fix atmospheric N biologically, store N in the soil, sequester C, minimizes C footprints of agriculture, enriches above and belowground biodiversity and enhances productivity of the cropping system. Further, legume-cereal intercropping system ensures better utilization of available resources and conserves soil and nutrients of top soil in erosion prone regions by covering maximum land area. Intercropping system provides natural insurance against crop failure and enables better ecosystem services and all these lead towards agricultural sustainability (Maitra *et al.*, 2019).

8.1 Intercropping, the Appropriate Option under Limited Resource Conditions

Intercropping may be considered as the appropriate option for sustainable crop production. Further, It follows the principal components of ecosystem services by facilitating diversity and complements among the crops grown together and ensures higher combined productivity. In intercropping, preferably dissimilar crops in terms of their morphology are selected with dissimilar duration which can ensure efficient utilization of limiting resources, namely, soil moisture, land, light and nutrients. As greater area of land is covered by the crops grown in association, soil moisture loss is checked under water-scarce situations in drylands and soil temperature is well maintained (Miao *et al.*, 2016). The mixture of deep and shallow rooted crops in intercropping is also beneficial as available soil nutrients are better used and moisture sharing is observed by the process of bio-irrigation. Moreover, greater coverage of land area restricts run-off of water, erosion of soil and loss of nutrients from topsoil. But the complement and yield advantage occur due to several factors, namely, choice of crops and varieties, maturity, seed rate and planting geometry and duration of the crops and system as a whole (Maitra *et al.*, 2020). Maximum advantage is obtained if all the factors are properly considered while planning for intercropping.

8.2 Potential of Legumes in Intercropping

The main objectives of intercropping are optimum use of resources available and achieving production sustainability. The cereal-legume combination of intercropping is an automatic choice to the smallholders in developing and tropical countries (Figure 8.1). In subsistence farming, the combination provides variety in the food basket and nutritional security. In developed temperate countries, forage cereal + legume mixed cropping is observed for nourishment of livestock (Figure 8.2). The cereal component is nutrient exhaustive in nature and absorbs nutrients from upper layers of the soil; however, some of the legumes are having deep root system (example, red gram and soybean) which can penetrate in deeper layers of the soil and absorb nutrients. Thus, the available nutrients of the soil are properly utilized without much competition among crop species grown in intercropping. Under limited water conditions also, two types of plants absorb moisture from different layers of the soil. The on-farm experiment carried out in Koraput of south Odisha indicated that mixed stands of finger millet +red gram and black gram in replacement series reduced the yield of finger millet than pure stand (Dass and Sudhishir, 2010). However, the combined yield of finger millet and legumes was more and both the combinations of finger millet with red gram and black gram produced more finger millet equivalent yield (FMEY) than sole finger millet. The resource use efficiency was greater in intercropping as the highest land equivalent ratio (LER) value of 1.34 was recorded from the intercropping of finger millet + pigeon pea with the row proportion of 5:2 and 6:2. In the study, intercropping finger millet and black gram covered maximum ground area and as a result the lowest runoff (10.2%) of water and nutrient loss through erosion was recorded over sole cropping of finger millet. when sown in contour because black gram covered enough ground area in intercropping with finger millet. The intercropping combination of finger millet and legumes also increased soil fertility status because legumes in association fixed nitrogen biologically and leaf fall of legume added organic matter to the marginal soil. In this way, the cereals and legumes combination of intercropping enhances soil fertility, checked runoff of water and nutrient loss from the top soil.

Figure 8.1. Cowpea (3 rows) intercropped in paired row maize (2 rows)

The complement among the crops chosen is one of the important factors of success of intercropping under limited resource conditions and it ensures optimum utilization of resources. In cereal – legume intercropping systems, if the crops are selected with different duration with the period for peak demand of nutrients, then complement is observed. Intercropping of maize (*Zea mays*) + mung (*Vigna radiata*) is the example of such complement where being a short duration crops mung expresses its peak nutrient requirement at 35 days after sowing, whereas the initial growth of maize is slow and its peak nutrient demand starts at the seventh week after sowing; by this time mung finishes a major part of the crop cycle (Ofori and Stern, 1987; Maitra *et al.*, 2019). Minimum level of competition among the component crops is also beneficial and it is common in cereal + legume intercropping systems. For example, in case of maize + soybean (*Glycine max*), both the cereal and legume are of same duration, but they differ in canopy height. Under this situation, maize uses more sunlight, water and nutrients from the common pool of available resources and legume uses water of deeper soil, fixes N biologically and fulfills its requirement and tolerates partial shade. An experiment carried out in Province Sichuan, China resulted in a positive impact of defoliation of top two maize leaves at silk stage and the effect was reflected as yield enhancement of maize and soybean intercropping system because of more light interception and dry matter partitioning (Raza *et al*., 2019). Therefore, less competition was observed between two species and there would be an increase in combined yield than pure stand. Earlier, Rezig *et al*. (2013) conducted a field trial in Tunisia on intercropping potato and French honeysuckle or *sulla* (*Hedysarum coronarium* L.), a crop of Fabaceae family and observed that pure stands of potato and *sulla* intercepted more photosynthetically active radiation (PAR) energy than in mixed stands, but a combined PAR interception per unit area was more in the intercropping combination. Generally, the difference in maturity of crops for a month gives a better opportunity to the longer duration crop component to use the resources alone after harvest of the short duration component. This competition free period can also result in more yields of the long duration crop and ultimately higher system productivity.

Figure 8.2. Fenugreek intercropped in sweet corn

Further, spatial arrangement with distinct row proportions and planting geometry also influences the success of intercropping. Paired row planting of base crop provides more space for intercrop in between two pair and if the intercrop is of short duration, after harvest of intercrop, base crop utilizes entire space to flourish (Maitra *et al*., 2020; Maitra *et al*., 2021; Panda *et al*., 2021). In Sundarbans of West Bengal, planting of paired row cotton registered significantly more productivity than uniform row cotton when both intercropped with green gram and groundnut and in the study paired row cotton + legumes resulted in greater monetary advantage than uniform row cotton (Maitra *et al*., 2001). Later, Kumar *et al*. (2018) also observed the benefits of paired row planting geometry in pearl millet and green gram intercropping system where intercropping legumes in paired row pearl millet yielded more compared to uniform row planting of pearl millet and intercropping. Yamuna *et al*. (2015) recorded that maize + grain legumes yield more from unit area in the southern dry zone of Karnataka.

8.3 Intercropping Legumes and Improvement of Soil Health

8.3.1 Soil Physical Properties

The smallholders prefer intercropping for their food and livelihood security. Generally, enough investment in agriculture is not observed in drylands where small and marginal farmers adopt farming for subsistence. Under this situation proper technological intervention for land and resource management is not followed. But inclusion of legumes in intercropping ensures soil management having influences on the physical soil properties. The positive impacts on some soil properties, namely, soil aggregation and infiltration of water due to proliferation of deep-rooted legumes were recorded in maize and legume intercropping systems (Latif *et al.*, 1992). The leaf-fall and residues of legume adds organic matter to the soil which improves soil structure (better soil aggregation), water holding capacity (Yadav *et al.*, 2017) and checks soil from disintegration. The residues of legumes have lower C:N ratio and they decompose faster and soil organic matter is enriched quickly (Layek *et al.*, 2018). When maize is intercropped with cowpea, the maximum land is covered and as a result soil erosion is checked (Kariag, 2004; Maitra *et al.*, 2020) and restricted run-off of water enables more infiltration of water. Ozturkmen *et al.* (2020) recorded that a mixed stand of cereal and legume enhanced aggregate stability and porosity of the soil. Xu *et al.* (2021) recorded that intercropping lowered soil bulk density than pure stand in their experiment carried out in southwest China. Yuvaraj *et al.* (2020) mentioned that intercropping legumes improved soil structures and added organic matters to the soil.

8.3.2 Soil Chemical Properties

In cereal – legume intercropping system, legumes play a vital role in nutrient uptake by crops than pure stand of cereals. In the soils with low available nitrogen, a combination of cereal and legume performed well (Vesterager *et al.*, 2008). In general, as the combined biomass yield is more in intercropping, nutrients uptake by the crops is more (Cardinale *et al.*, 2007). In cereal + legume intercropping, not only nitrogen, but phosphorus and potassium uptake in intercrops were also more in cereal – legume intercropping system compared to pure stands (Morris and Garrity, 1993). An increased phosphorus uptake was reported by the scientists under various intercropping system with legumes chosen in combination with cereals. For example, pigeon pea + sorghum intercropping (Ae *et al.*, 1990) and lupin + wheat intercropping (Cu *et al.*, 2005). In acid soils, Al toxicity was reduced (Ryan *et al.*, 2011). The micro-biome plays an important role in availability of soil nutrients to crops more prominently in cereal-legume intercropping system. Several legume salter soil pH (Willey and Osiru, 1972; Mc Lay, 1997; Sas, 2001; Cheng, 2004; Nasar *et al.*, 2019) and make P, K, Ca, and Mg more available to plants (Dakora, 2003). Legumes lower rhizospheric soil pH by releasing organic anions and increase the availability of organic P (Kamh, 1999; Li, 2004). Hailu and Geremu (2021) noted that intercropped sorghum and legumes resulted in higher C:N ratio and cation exchange capacity (CEC) in eastern Ethiopia. Nwite *et al.* (2017) recorded that soil pH was increased from strongly acidic to slightly acidic due to intercropping maize and legumes in Nigeria. Further, they recorded not only nitrogen, but also calcium, organic carbon and magnesium were enhanced as a result of intercropping legumes in maize. Moreover, pigeon pea + sorghum intercropping system increased P uptake by crops (Muofhe and Dakora, 2000). In a recent study, Malviya *et al.* (2021) recorded that intercropping sugarcane + soybean reduced the soil pH slightly, but enhanced total and available soil N, P and K status in Guangxi, China. The intercropping of finger millet + black gram grown in contour resulted in the reduced runoff of water and soil loss and N, P and K loss through erosion of topsoil (Dass and Sudhishir, 2010). Actually, selection of appropriate crops species and their proportion ensures the mutual benefits in cereal-legume intercropping system and thus, the enhanced chemical and biological activities in the soil can assure availability of some micro-nutrients like Fe and Zn (Xue *et al.*, 2016).

8.3.3 Legumes in Intercropping and N Transfer

Legumes are known to fix nitrogen biologically and in association with non-legumes, they share the fixed N beyond fulfillment of their own requirement to non-legumes and thus, non-legumes get benefit. The N-transfer to non-legumes occurs in the form of root excretion and leaf fall of legumes. Eaglesham *et al.* (1981) reported that in cowpea + maize intercropping, about 25% of N fixed by cowpea was shared with maize. In an experiment, *faba* bean contributed 15% N needs of wheat in intercropping (Xiao *et al.*, 2004). The N transfer from groundnut to aerobic rice ranged between 5.5 to 11.4% percent under graded level of nitrogenous fertilizer application (Chu *et al.*, 2004). Worldwide, maize-legume intercropping is very common and it was noted that contribution of green gram was 7-11% of N fixed by legume, however, cowpea and groundnut shared 11-20% and 12-26% of biologically fixed N by them to maize (Senaratne *et al.*, 1995). In sorghum + soybean intercropping system, legume component shared 7% of N to cereal (Ofosubudu *et al.*, 1993), but in another study, it ranged between 20 to 55 percent (Ofosubudu *et al.*, 1995).

8.3.4 Soil Biological Properties

Legumes not only enrich soil by adding N fixed from the atmosphere but also influence in changing the diversity of beneficial micro-organisms in the rhizosphere and thus improve soil biological properties. The associated non-legume crop species in intercropping with legumes gets the benefit. Due to fixation of N by legumes in intercropping, yields of cereal as well as total productivity are increased as observed in sorghum+ soybean (Fujita *et al.*, 1992), maize + cowpea (Kimou *et al.*, 2017), maize + groundnut/ black gram (Manasa *et al.*, 2018; Manasa *et al.*, 2020) and so on. The higher yields in the intercropping system are noted due to complement among the crop species and choice of the crops with inherent capability to use resources efficiently (Willey, 1979). Interestingly, in soils with poor N content, Maize + cowpea intercropping system yielded more (Vesterager *et al.*, 2008) and cereals performed well due to N transfer by legumes (Herridge *et al.*, 1995). Legumes can reduce the rhizospheric soil pH by releasing organic anions (Eaglesham, 1981; Li, 2003; Li, 2004) and enhance the organic P availability to plants and soil microorganisms. The leguminous crops fix N biologically and transfer a potion to non-legume (cereals) in intercropping and it was noted that this N transfer increased the ability of legumes to fix more N and thus, efficient N fixation and utilization by crops resulted in higher system productivity (Dinkelaker, 1989; Frey and Schüepp, 1993).

Legumes improve soil microbial population (Meena *et al.*, 2014), enrich soil with low molecular weight organic compound as exudates (Layek *et al.*, 2018) and enzymatic activities such as phosphatase, dehydrogenase, protease, urease, and β-glucosidase in the rhizosphere (Roldan *et al.*, 2003). The population of other beneficial micro-organisms like *Pseudomonas* sp., Cyanobacteria, Alphaproteo-bacteria, and Betaproteo-bacteria are also increased (Spehn *et al.* 2000; Qiao *et al.*, 2012; Li and Wu, 2018). The enhancement of beneficial microbial population is synonymous to the build up of a healthy ecosystem (Maitra and Ray, 2019).

Legumes are known to increase soil C and reduce C:N ratio which facilitates higher microbial activities and when legumes are intercropped with cereals, automatically the soil becomes biologically active than sole cropping of cereal (Nasar *et al.*, 2019). In addition to N, under P deficient soil conditions, P uptake by crops is improved in cereal + legume intercropping (Tarafdar and Jungk, 1987; Muofhe and Dakora, 2000) due to more phosphatase activities. Chickpea releases rhizospheric acid and phosphatase and *faba* bean releases malate, protons and citrates into the rhizosphere that mobilize P and associated cereals in intercropping get the benefit (Nasar *et al.*, 2019). In legume-based intercropping, crops can perform well because of increased availability of P by legume factor, but under low P conditions N fixation by legumes is decreased (Eivazi and Tabatabai, 1977; Nakas, 1987).

Further, in wheat + chickpea intercropping release of hydroxyl ions raised soil pH and P availability was more in a P-deficient neutral soil (Betencourt *et al.*,

2012). Enhancement of micro-biological diversity in legume-based intercropping systems increase complexity in below ground soil environments. The P availability in the soil is altered by root-mediated and microbe-mediated processes in legume + cereal intercropping system (Xue *et al.*, 2016). Choudhary and Choudhary (2016) noted enhanced P and K balance by maize-legume intercropping. Further, *Rhizobium* species suppress growth of pathogens like *Fusarium* spp., *Phytophthora* spp. and so on (Tu, 1978; Buonassisi *et al.*, 1986; Omar *et al.*, 1998). Malviya *et al.* (2021) recorded that in sugarcane + legume intercropping system, the micro-organisms population of *Acidobacteria*, *Chloroflexi*, *Cyanobacteria*, *Verrucomicrobia*, *Armatimonadetes* and *Nitrospirae* were predominant and their population was more compared to sole cropping of sugarcane. Among fungal phyla *Zygomycota*, *Ascomycota*, *Basidiomycota*, *Chytridiomycota* and *Glomoromycota* dominated in the rhizosphere in the intercropped field.

The impact of legumes as a component in intercropping is observed for enhancing crop yields, maintaining soil health and sustaining agriculture under all ecological conditions and more particularly under fragile and resource poor conditions. The non-legume or cereal component is benefited much in association with legumes in intercropping. Legumes have enough potential to mitigate different issues related to climate change and global warming. Hence, inclusion of legume in sequential as well as in intercropping system may be considered as the right step to attain agricultural sustainability.

References

Ae, N., Arihara, J., Okada, K., Yoshihara, T., Johansen, C. 1990. Phosphorus uptake by pigeon pea and its role in cropping systems of the Indian subcontinent. *Sci.* **248**:477-480.

Bedoussac, L., Journet, E. P., Hauggaard-Nielsen, H., Naudin, C., Corre-Hellou, G., Jensen, E. S., Prieur, L. and Justes, E. 2015. Ecological principles underlying the increase of productivity achieved by cereal-grain legume intercrops in organic farming: a review. *Agron Sustain Dev.* **35**:911–935.

Betencourt, E., Duputel, M., Colomb, B., Desclaux, D. and Hinsinger, P. 2012. Intercropping promotes the ability of durum wheat and chickpea to increase rhizosphere phosphorus availability in a low P soil. *Soil Biol. Biochem.* **46**:181–190.

Buonassisi, A. J., Copeman, R. J., Pepin, H. S. and Eaton, G. W. 1986. Effect of Rhizobium spp. on *Fusariumsolani* f. sp. *phaseoli. Can. J. Plant Pathol.* **8**:140–146.

Cardinale, B. J., Wright, J. P., Cadotte, M. W., Carroll, I. T., Hector, A., Srivastava, D. S., Loreau, M. and Weis, J. J. 2007. Impacts of diversity on biomass production increase through time because of species complementarity. *Proc. Nat. Acad. Sci.*, USA. **104**:18123-18128.

Cheng, Y. 2004. Proton release by roots of Medicago murex and Medicago sativa growing in acidic conditions, and implications for rhizosphere pH changes and nodulation at low pH. *Soil Biol. Biochem.* **36**(8):1357-1365.

Choudhary, V. K. and Chaoudhary, B.U. 2016. A staggered maize-legume intercrop arrangement influences yield, weed smothering and nutrient balance in the Eastern Himalayan Region of India. *Exp. Agric.* **54(2):** 181-200. DOI: 10.1017/S0014479716000144.

Chu, G. X., Shen, Q, R. and Cao, J. L. 2004. Nitrogen fixation and N transfer from peanut to rice cultivated in aerobic soil in an intercropping system and its effect on soil N fertility. *Plant Soil*. **263**: 17–27.

Cu, S. T., Hutson, J. and Schuller, K. A. 2005. Mixed culture of wheat (*Triticum aestivum* L.) with white lupin (*Lupin usalbus* L.) improves the growth and phosphorus nutrition of the wheat. *Plant Soil*. **272**(1- 2):143-151.

Dakora, F. D. 2003. Defining new roles for plant and rhizobial molecules in sole and mixed plant cultures involving symbiotic legumes. *New Phytol*. **158**:39-49.

Dass, A. and Sudhishir, S. 2010. Intercropping in finger millet (*Eleusine coracana*) with pulses for enhanced productivity, resource conservation and soil fertility in uplands of Southern Orissa. *Ind. J. Agron.* **55** (2):89-94.

Dinkelaker, B. 1989. Citric acid excretion and precipitation of calcium citrate in the rhizosphere of white lupin (*Lupinus albus* L.). *Plant Cell Environ.* **12**(3):285-292.

Eaglesham, A. 1981. Improving the nitrogen nutrition of maize by intercropping with cowpea. *Soil Biol. Biochem*. **13(2)**:169-171.

Eaglesham, A. R. J., Ayanaba, A., Ranga Rao, V. and Eskew, D. L. 1981. Improving the nitrogen nutrition of maize by intercropping with cowpea. *Soil Biol. Biochem*. **13**:169-171.

Eivazi, F. and Tabatabai, M. 1977. Phosphatases in soils. *Soil Biol. Biochem*. **9**(3): 167-172.

Frey, B. and Schüepp, H. 1993. A role of vesicular-arbuscular (VA) mycorrhizal fungi in facilitating interplant nitrogen transfer. *Soil Biol. Biochem*. **25**(6): 651-658.

Fujita, K., Ofosu-Budu, K. G. and Ogata, S. 1992. Biological nitrogen fixation in mixed legume-cereal cropping systems. *Plant Soil*. **141**:155–176.

Hailu, H and Geremu, T. 2021. Effect of sorghum-legume intercropping patterns on selected soil chemical properties and yield of sorghum at midland areas of west hararghe zone of oromia regional state, eastern ethiopia. *Irrig. Drain. Sys. Eng.* **10**(5):1-8.

Herridge, D. F., Maecellos, H., Felton, W. L., Turner, G. L. and Peoples, M. B. 1995. Chickpea increases soil N fertility in cereal systems through nitrate sparing and N_2 fixation. *Soil Biol. Biochem*. **27**:545–551.

Kamh, M. 1999. Mobilization of soil and fertilizer phosphate by cover crops. *Plant Soil*. **211**: 19.

Kariag, B. M. 2004. Intercropping maize with cowpeas and beans for soil and water management in Western Kenya. In: Proc. of the 13th International Soil Conservation Organization Conference, July 4-9, 2004. Brisbane: Conserving Soil and Water for Society, pp. 1-5. DOI: 2010.135.145.

Kimou, S. H., Coulibaly, L. F., Koffi, B. Y., Toure, Y., Dede, K. J. and Kone, M. 2017. Effect of row spatial arrangements on agro morphological responses of maize (*Zea mays* L.) and cowpea *(Vigna unguiculata* L.). *Afr. J. Agric. Res.***12**(34):2633-2641. DOI: 10.5897/AJAR2017.12509.

Kumar, V., Singh, R. P., Kumar, S., Shukla, U. N. and Kumar, K. 2018. Performance of pearlmillet + greengram intercropping as influenced by different planting techniques and integrated nitrogen management under rainfed condition. *Int. J. Chem. Stud.***6**(3): 705-708.

Layek, J., Das, A., Mitran, T., Nath, C., Meena, R. S., Yadav, G. S., Shivakumar, B. G., Kumar, S. and Lal, R. 2018. Cereal + Legume Intercropping: An Option for Improving Productivity and Sustaining Soil Health, *In:* R. S. Meena *et al.* (*Eds.*), Legumes for Soil Health and Sustainable Management, Springer Nature Singapore Pte Ltd., pp. 347-386. DOI: 10.1007/978-981-13-0253-4_11.

Li, S. and Wu, F. 2018. Diversity and Co-occurrence Patterns of Soil Bacterial and Fungal Communities in Seven Intercropping Systems. *Front. Microbiol*. **9**:15-21.

Li, S. M. 2004. Acid phosphatase role in chickpea/maize intercropping. *Ann. Bot*. **94**(2): 297-303.

Li, W. 2003. Effects of nitrogen and phosphorus fertilizers and intercropping on uptake of nitrogen and phosphorus by wheat, maize, and faba bean. *J. Plant. Nutr*. **26**(3): 629-642.

Latif, M. A., Mehuys, G. R., Mackenzie, A. F., Alli, I., and Faris, M. A. 1992. Effects of legumes on soil physical quality in a maize crop. *Plant Soil.* **140**(1), 15-23.

Maitra, S.and Ray, D. P. 2019. Enrichment of biodiversity, influence in microbial population dynamics of soil and nutrient utilization in cereal-legume intercropping systems: A Review. *Int. J. Bioresour. Sci*. **6**(1):11-19. DOI: 10.30954/2347-9655.01.2019.3.

Maitra, S., Palai J. B., Manasa, P. and Prasanna Kumar, D. 2019. Potential of intercropping system in sustaining crop productivity. *Int. J. Agric. Environ. Biotechnol.* **12**(1):39-45. DOI: 10.30954/0974-1712.03.2019.7.

Maitra, S.,Samui, R. C**.,** Roy, D. K. and Mondal, A. K. 2001**.** Effect of cotton based intercropping system under rainfed conditions in *Sundarban* region of West Bengal. *Ind. Agric*. **45 (**3-4**):** 157-162.

Maitra, S., Shankar, T. and Banerjee, P. 2020. Potential and Advantages of Maize-Legume Intercropping System (Online First), Intech Open, (*In:Maize - Production and Use, Ed.* Akbar Hossain), DOI: 10.5772/intechopen.91722. Available at: www. intechopen.com (Accessed on 26 March, 2020).

Maitra, S., Hossain, A., Brestic, M., Skalicky, M., Ondrisik, P., Gitari, H., Brahmachari, K., Shankar, T., Bhadra, P., Palai, J.B., Jana, J., Bhattacharya, U., Duvvada, S. K., Lalichetti, S., Sairam, M. 2021. Intercropping– a low input agricultural strategy for food and environmental security. *Agronomy*. **11**: 343. https://doi.org/10.3390/ agronomy 11020343.

Malviya, M. K., Solanki, M. K., Li, C. N., Wang, Z., Zeng, Y., Verma, K. K., Singh, R. K., Singh, P., Huang, H. R., Yang, L. T., Song, X. P. and Li, Y. R. 2021. Sugarcane-legume intercropping can enrich the soil microbiome and plant growth. *Front. Sustain. Food Syst*. **5**:606595. doi: 10.3389/fsufs.2021.606595.

Manasa, P., Maitra, S. and Barman, S. 2020. Yield attributes, yield, competitive ability and economics of summer maize-legume intercropping system. *Int. J. Agric. Environ. Biotechnol* **13**(1): 33-38, DOI: 10.30954/0974-1712.1.2020.16.

Manasa, P., Maitra, S. and Reddy, M. D. 2018. Effect of Summer Maize-Legume Intercropping System on Growth, Productivity and Competitive Ability of Crops. *Int. J. Eng. Res. Manag. Technol*. **8**(12), 2871-2875.

Mc Lay, C. 1997. Acidification potential of ten grain legume species grown in nutrient solution. *Aus. J. Agric. Res*. **48**:1025-1032.

Meena, V. S., Maurya, B. R., Meena, R. S., Meena, S. K., Singh, N. P. and Malik, V. K. 2014. Microbial dynamics as influenced by concentrate manure and inorganic fertilizer in alluvium soil of Varanasi, India. *Afr. J. Microb. Res*. **8**(1):257–263.

Miao, Q., Rosa, R. D., Shi, H., Paredes, P., Zhu, L., Dai, J., Gonçalves, J. M. and Pereira, L. S. 2016. Modeling water use, transpiration and soil evaporation of spring wheat–maize and spring wheat–sunflower relay intercropping using the dual crop coefficient approach. *Agric. Water Manage*. **165**:211–229.

Morris, R. A. and Garrity, D. P. 1993. Resource capture and utilization in intercropping; non-nitrogen nutrients. *Field Crops Res.***34**(3-4):319-334.

Muofhe, M. L. and Dakora, F. D. 2000. Modification of rhizosphere pH by the symbiotic legume Aspalathus linearis growing in a sandy acidic soil. *Funct. Plant Biol.***27**(12):1169-1173.

Yuvaraj, M., Pandiyan, M. and Gayathri, P. 2020. Role of Legumes in Improving Soil Fertility Status, Legume Crops - Prospects, Production and Uses, Mirza Hasanuzzaman, IntechOpen, DOI: 10.5772/intechopen.93247. Available from: https://www.intechopen.com/chapters/72818.

Nakas, J. 1987. Origin and expression of phosphatase activity in a semi-arid grassland soil. *Soil Biol. Biochem*. **19**(1):13-18.

Nasar, J., Alam, A., Nasar, A. and Khan, M. Z. 2019. Intercropping induces changes in above and below ground plant compartments in mixed cropping system. *Biomed. J. Sci. Tech. Res*. **17**(5):13043-13050, DOI: 10.26717/BJSTR.2019.17.003054.

Nwite, J. N., Njoku, C. and Alu, M.O. 2017. Effects of intercropped legumes with maize (Zea mays L.) on chemical properties of soil and grain yield of maize in Abakaliki, Nigeria. *Niger. Agric. J.***48**(2): 105-112.

Ofori, F. and Stern, W. R. 1987. Cereal-legume intercropping systems. *Adv. Agron.* **41**:41–90.

Ofosubudu, G. K., Sumiyoshi, D., Matsuura, H. and Fujita, K. 1993. Significance of soil N on dry-matter production and N-balance in soybean sorghum mixed cropping system. *Soil Sci. Plant Nutri.* **39**: 33–42.

Ofosubudu, K. G., Noumura, K. and Fujita, K. 1995. N-2 fixation, N transfer and biomass production of soybean cv bragg or its supernodulating nts1007 and sorghum mixed-cropping at 2 rates of N fertilizer. *Soil Biol. Biochem.* **27**: 311–317.

Omar, S.A. and Abd-Alla, M.H. 1998. Bio-control of fungal root rot diseases of crop plants by the use of Rhizobia and Bradyrhizobia. *Folia Microbiol.* **43**:431–437.

Ozturkmen, A. R., Ramazanoglu, E., Almaca, A. and Çakmakli, M. 2020. Effect of intercropping on soil physical and chemical properties in an olive orchard. *Appl. Ecol. Environ. Res.* **18**(6):7783-7793.

Panda, S. K., Maitra, S., Panda, P., Shankar, T., Pal, A., Sairam, M. and Praharaj, S. 2021. Productivity and competitive ability of rabi maize and legumes intercropping system. *Crop Res.* **56** (3 & 4):98-104; DOI:10.31830/2454-1761.2021.016.

Qiao, Y. J., Li, Z. Z., Wang, X., Zhu, B., Hu, Y. G. and Zeng, Z. H. 2012. Effect of legume-cereal mixtures on the diversity of bacterial communities in the rhizosphere. *Plant Soil Environ.* **58**(4):174–180.

Raza, M. A., Feng, L. Y., Van der Wref, W., Iqbal, N., Khan, I., Hassan, M. J., Ansar, M., Chen, Y. K., Xi, Z. J., Shi, J. Y., Ahmed, M., Yang, F. and Yang, W. 2019. Optimum leaf defoliation: a new approach for increasing nutrient uptake and land equivalent ratio of maize soybean relay intercropping system. *Field Crop Res.***244**:107647. DOI:10.1016/j.fcr.2019.107647.

Rezig, M., Sahli, A, Hachicha, M., Jeddi B.F. and Harbaoui, Y. 2013. Light interception and radiation use efficiency from a field of potato (Solanum tuberosum L.) and sulla (Hedysarumcoronarium L) intercroppinh in Tunisia. *Asian J. Crop Sci.***5**(4): 378-392.

Roldan, A., Caravaca F, Hernandez, M. T., Garci, C., Sanchez-Brito, C., Velasquez, M. and Tiscareno, M. 2003. No-tillage, crop residue additions, and legume cover cropping effects on soil quality characteristics under maize in Patzcuaro watershed (Mexico). *Soil Tillage Res.* **72**:65–73.

Ryan, P. R., Tyerman, S. D., Sasaki, T., Furuichi, T., Yamamoto, Y., Zhang, W. H. and Delhaize, E. 2011. The identification of aluminium-resistance genes provides opportunities for enhancing crop production on acid soils. *J. Exp. Bot.***62**: 9–20.

Sas, L. 2001. Excess cation uptake, and extrusion of protons and organic acid anions by Lupinus albus under phosphorus deficiency. *Plant Sci.***160**(6):1191-1198.

Senaratne, R., Liyanage, N. D. L. and Soper, R. J. 1995. Nitrogen-fixation of and N-transfer from cowpea, mungbean and groundnut when intercropped with maize. *Fertil. Res.* **40**: 41–48.

Spehn, E. M., Joshi J., Schmid, B., Alphei, J. and Korner, C. 2000. Plant diversity effects on soil heterotrophic activity in experimental grassland ecosystems. *Plant Soil.* **224**:217–230.

Tarafdar, J. and Jungk, A. 1987. Phosphatase activity in the rhizosphere and its relation to the depletion of soil organic phosphorus. *Biol. Fertil. Soils.* **3**(4): 199-204.

Tu, J.C. 1978. Protection of soybean from severe *Phytophthora* root rot by *Rhizobium. Physiol. Plant Pathol.* **12**: 233–240.

United Nations, 2019. Growing at a slower pace, world population is expected to reach 9.7 billion in 2050 and could peak at nearly 11 billion around 2100, available at: www.un.org/development/desa/en/news/ population/world-population-prospects-2019.html (accessed on: 14 April, 2020).

Vesterager, J. M., Nielsen, N. E. and Høgh-Jensen, H. 2008. Effects of cropping history and phosphorus source on yield and nitrogen fixation in sole and intercropped cowpea-maize systems. *Nutr. Cycl. Agroecosyst.* **80**:61-73. DOI: 10.1007/ s10705-007-9121-7.

Willey, R. W. 1979. Intercropping-Its importance and research needs. Part 1: Competition and yield advantages. *Field Crops Res.* **32**:1-10.

Willey, R. W. and Osiru, D. 1972. Studies on mixtures of maize and beans (*Phaseolus vulgaris*) with particular reference to plant population. *J. Agric. Sci.* **79**: 517-529.

Xiao, Y. B., Li, L. and Zhang, F. S. 2004. Effect of root contact on interspecific competition and N transfer between wheat and fababean using direct and indirect N-15 techniques. *Plant Soil,* **262**: 45–54.

Xu, Q., Xiong, K., Chi, Y. and Song, S. 2021. Effects of crop and grass intercropping on the soil environment in the karst area. *Sustain.* **13**:5484. https://doi.org/10.3390/su13105484.

Xue, Y., Xia, H., Christie, P., Zhang, Z., Li, L. and Tang, C. 2016. Crop acquisition of phosphorus, iron and zinc from soil in cereal/legume intercropping systems: a critical review. *Ann. Bot.* **117**:363–377.

Yadav, G. S., Lal, R., Meena, R. S., Babu, S., Das, A., Bhowmik, S. N. and Saha, P. 2017. Conservation tillage and nutrient management effects on productivity and soil carbon sequestration under double cropping of rice in north eastern region of India. *Ecol. Indic.* **105**: 303-315.

Yamuna, B. G., Yogananda, S. B., Denesh, G. R., Prashanth Kumar, M. K. and T. N. Ranjini, T. N. 2015. Growth and yield of maize as influenced by maize based intercropping system for southern dry zone of Karnataka. *Ecoscan*, **7**: 335-339.

9

Performance of Crops in Intercropping

Intercropping with food crops and vegetables is widely practiced by small and marginal farmers of the world mainly belonging to arid and semi-arid regions. Because of several benefits obtained, farmers prefer to adopt intercropping system. In temperate regions, mixed stands of cover crops and combinations of cereal and legumes forage production are very common. But in the tropics and semi-arid tropics, poor farmers try a lot to ascertain food and nutritional security where intercropping actually exerts its significance (Behera, 2017). Crop choice, seeding proportion and planting geometry, and management of crops in mixed stands are the key considerations for the success of the intercropping system. Research carried out in several parts of the world indicated promising results in favour of intercropping. Based on earlier research evidences, the chapter is narrated highlighting positive impacts of intercropping on cultivation of important cereals and millets.

9.1 Maize-based Intercropping System

Maize (*Zea mays* L.) belongs to the family *Poaceae* and originated in Mexico. Maize is an important cereal ensures food security to the vast population in developing countries. Maize is not only used as food but as animal feed and also became essential in food processing industries and which made it popular in the world. Hence, it is called 'Queen of Cereals'. The nutrient content of maize is with 11.1% protein, 66.3% carbohydrate, 3.6% fat, 2.7% fiber and 1.5% minerals such as calcium, phosphorus, iron and vitamins (A, B, E) (Joshi *et al.,* 2017). The quality protein maize (QPM) cultivars developed aiming to alleviate malnutrition by enriching protein content in maize (Tripathy, 2019). Maize is grown in an exceedingly large selection area of the world, starting from the temperate hill zones to semi-arid desert conditions in all the three cultivable seasons. In the world scenario of maize, the production was 1148.5 million t from 197.2 million ha with an average yield of 5.82 t/ha (FAOSTAT, 2022). In India, the total area cultivated under maize was 9.18 million ha with a production and average productivity of 27.23 million t and 2965 kg/ha, respectively during 2018-19 (ICAR-IIMR 2019-20). As the crop can be grown irrespective of season with wider adaptability, it provides enough scope to cultivate any time during the year. Besides, the row spacing maintained in maize is fairly wide (60 cm and above) that provides the scope of accommodating other crops in mixed stands. Preferably, legumes are intercropped by smallholders for food and nutritional security. The combination of maize and legume was studied worldwide in the intercropping system and benefits were addressed by the researchers (Manasa *et al.,* 2018; Panda *et al.,* 2021). The following paragraphs narrated the benefits of intercropping maize with other crops.

9.1.1 Yield Advantages

The row spacing of maize is wide and it is treated as base-crop where 100% population of maize is maintained in intercropping with other crops even in uniform row planting method. However, in paired row planting of maize, additional space is created for accommodating intercrops. Legumes are the most compatible crop with maize in intercropping. The land equivalent ratio (LER) is one of the key indicators of yield advantage and resource use efficiency in an intercropping system. The LER denotes the proportion of yield of crops grown in an intercropping system in relation to the yields of corresponding crops under sole cropping. Some of the promising LER values have been presented under Table 9.1 based on earlier studies. Actually, the LER specifies the advantages of intercropping (Mead and Willey, 1980) and the value of LER more than 1.0 is suggestive to yield advantages with greater land use efficiency.

Table 9.1. LER values of some maize-based intercropping systems

Intercropping system	Proportion	LER	Country	Reference
Maize + groundnut	2:2	1.42	Ghana	Kermah *et al.* (2017)
Maize + garden pea	1:2	1.56	Bangladesh	Khan *et al.* (2018)
Maize + groundnut	2:2	1.82	India	Manasa *et al.* (2018)
Maize + soybean	2:2	1.90	China	Raza *et al.* (2019)
Maize + groundnut	2:1	1.75	India	Panda *et al.* (2021)
Maize + groundnut	2:3	1.74	India	Panda *et al.* (2021)

The area time equivalent ratio (ATER) further fine-tunes the LER with the time factor for which the crops occupied the land area (Hiebsch, 1978). The ATER value with more than unity also indicates benefits of an intercropping system. Earlier researchers also recorded the advantageous ATER of maize + legume intercropping system (Table 9.2).

Table 9.2. ATER values of some maize-based intercropping systems

Intercropping system	Proportion	ATER	Country	Reference
Maize + black gram	1:2	1.47	India	Choudhary (2014)
Maize + soybean	2:6	1.32	India	Yogesh *et al.* (2014)
Maize + black cowpea	2:2	1.51	India	Jan *et al.* (2016)
Maize + groundnut	2:2	1.70	India	Manasa *et al.* (2018)
Maize + groundnut	2:3	1.49	India	Panda *et al.* (2021)

Maize equivalent yield (MEY) is another expression which is used to measure the efficiency of maize-based intercropping system over pure stand of maize.

Table 9.3. Maize-equivalent yield (MEY) in maize and legume intercropping system

Intercropping system	Ratio	MEY (kg/ha)	Sole maize yield (kg/ha)	Sole legume yield (kg/ha)	References
Maize + soybean	2:6	9470	7092	5450	Yogesh *et al.* (2014)
Maize + black cowpea	2:2	7699	5062	4785	Jan *et al.* (2016)
Maize + garden pea	1:2	20220	8200	6450	Khan *et al.* (2018)
Maize + groundnut	2:2	7609	5610	522	Manasa *et al.* (2019)

9.1.2 Other Benefits

The above-mentioned competitive functions indicated that maize in intercropping with other crops (Figure 9.1) performed well in terms of yield enhancement of crops from a unit area. Moreover, higher yield of crops was obtained because of efficient resource utilization (Willey, 1979).

Figure 9.1: Maize and chickpea intercropping system; **(A)** paired row maize + 3 rows chickpea (2:3) and **(B)** uniform row maize + single row chickpea (1:1)

The complementarity among the component crops made it possible (Willey and Reddy, 1981). Even with adoption of proper planting geometry and seeding proportion, component crops individually can yield higher than their respective pure stands as noted by Kimou *et al.* (2017) in maize + cowpea intercropping.

In the soils with low available N, mixed stands of maize and legumes yielded more (Vesterager *et al.,* 2008). Moreover, Ghanbari *et al.* (2010) recorded that intercropping maize + cowpea registered more light interception and restricted soil moisture evaporation. Other researchers also recorded efficient light utilization in maize-based intercropping system (Keating and Carberry, 1993; Jiao *et al.*, 2008). Further, the combination of shallow-rooted maize and deep-rooted legumes can utilize available soil moisture from different soil layers (Willey, 1979; Maitra *et al.*, 2020).

9.2 Sorghum-based Intercropping

Sorghum (*Sorghum bicolor* L. Moench) is a prime coarse-cereal of arid and semi-arid tropics of Asia and Africa. The grains of sorghum are widely used as food, but as feed and forage also it has significant demand. The other products prepared from sorghum are alcohol, bio-fuel, value-added foods, packaging materials etc. The recent development of sweet sorghum cultivars made it diversely used food and livestock feed (Arshad and Ranamukhaarachchi, 2012). The nutritional value of sorghum is quite healthy as each 100 g grain is consisting of 10.62 g protein, 72 g carbohydrate, 6.7 g fiber, 1.43 g ash and 3.45 g fat with about 330 KCal energy (USDA, 2019). Worldwide, sorghum is grown in more than 42 million ha and world production is about 60 million t (FAOSTAT, 2018). The USA produces maximum sorghum and India is considered as a leading sorghum growing country the world ranking in fifth position. The country produces about 5.0 million t of sorghum grain annually from around 5.0 million ha of cultivated area (GOI, 2021). In India, sorghum is mainly cultivated in resource-poor areas by small and marginal farmers mostly under rainfed conditions. As sorghum is an ecologically hardy crop, it performs fairly well under extreme climatic condition and weather aberrations. Intercropping provides an added advantage to smallholders by ensuring natural insurance and food security. Thus, sorghum is considered a coarse-cereal that facilitates the scope for intercropping with other crops because of diversity in food-grains basket of small farmers and in this regard, legumes as components of intercropping systems harness multifaceted benefits (Duvvada and Maitra, 2020).

9.2.1 Yield Advantages

Yield advantage is one of the important benefits obtained in the intercropping system and it is obtained because of efficient use of land and other resources. The land use efficiency can easily be quantified by LER and based on earlier researches in different countries of the world, the LER values of some experiments have been presented in Table 9.4.

Table 9.4. LER values of some sorghum-based intercropping

Intercropping systems	Row ratios	Combined LER values	Country	Reference
Sorghum + lima bean	5:1	1.26	Nigeria	Reza, 2012
Sorghum + soybean	1:1	1.40	Nigeria	Alfa, 2015
Sorghum + groundnut	1:1	1.67	Ethiopia	Dereje *et al.*,2016
Sorghum + cowpea	1:1	1.25	Ethiopia	Gebremichael *et al.*, 2019
Sweet sorghum + green gram	1:1	1.80	Pakistan	Arshad *et al.*, 2020
Sorghum + green gram	2:1	1.31	India	Chaithanya *et al.*, 2021
Sweet sorghum + cowpea	-	2.69	South Africa	Baker *et al.*, 2021
Sorghum + *lablab* bean	1:1	1.87	Ethiopia	Kahsay *et al.*, 2021

Similarly, there are some research works that evidenced benefits in terms of ATER in sorghum-based intercropping systems (Table 9.5).

Table 9.5. ATER values of some sorghum-based intercropping

Intercropping system	Proportion	ATER	Country	Reference
Sorghum + pearl millet	4:4	1.15	Pakistan	Amanullah *et al.*, 2020
Sorghum + black gram	2:1	1.18	India	Somu *et al.*, 2020
Sorghum + haricot bean	1:2	1.13	Ethiopia	Teferi *et al.*, 2020
Sorghum + red gram	2:1	1.25	India	Mallikarjun *et al.*, 2018

9.2.2 Other Benefits

The benefit of yield increase in intercropping is obtained because of efficient resource use (Maitra *et al.*, 2021). Sorghum cultivated as grain and forage registered advantages of intercropping. Intercropping sorghum and legumes is regarded to increase nutrient use efficiency (NUE) and soil fertility (Crews and Peoples, 2004). Forage sorghum grown in association with forage legumes (green gram, cluster bean, cowpea and *Dhaincha*) utilized the resources efficiently (Ahmad *et al.*, 2006). Intercropping red gram in sorghum increased P uptake by crops (Ae *et al.*, 1990). Greater amount of nitrogen, phosphorus and potassium uptake by intercropping of sweet sorghum + *Phaseolus trilobus* was noted by Singh *et al.* (2020). Similarly, Tripathi and Kushwaha (2014) more macro nutrients uptake by sorghum and pigeon pea intercropping systems. Other than N-fixation on intercropping legumes with sorghum, the system increases P availability (Crème *et al.*, 2016), decreases nutrient loss from the topsoil (Cavagnaro *et al.*, 2015), supports phyto-remediation (Chen *et al.*, 2015), and increases presence of beneficial microbes into the soil (Wahbi *et al.*, 2016) and soil organic carbon status (Tanwar *et al.*, 2014). Efficient weed management is also another benefit in sorghum based intercropping systems (Khan *et al.*, 2007; Machado, 2009;

Satheeshkumar *et al.* 2011). Further, functional diversity is created in an intercropping system that can reduce insects and diseases into the crop field. Research evidences supported that intercropping systems of various crops in sorghum reduce pests and diseases for sorghum and intercrops (Nickel, 1973; Litsinger and Moody, 1976; Narayanaswamy *et al.*, 1988; Maitra and Ray, 2019).

9.3 Pearl Millet in Intercropping System

Pearl millet is a coarse grain crop that is also known as poor man's meal. Since prehistoric times, it has been widely grown throughout Africa and Asia. In the comparatively arid areas of the world, it supplies basic food for the needy people (small and marginal farmers) in a short period of time. Among cereals and millets, it is the most drought-tolerant crop. The pearl millet grain has a high nutritional value with 12.4% moisture, 11.6% protein, 5% fat, 67% carbohydrates, and 2.7% minerals. It is also rich in iron, zinc, magnesium and phosphorus. It has higher contents of vitamin A, B_1, and B_6. Thus, it contributes to nutritional security along with food security in the areas that suffer from abiotic stresses. It is also utilized as poultry feed and forage for the cattle. India is the world's leading producer of pearl millet, with a 7.52-million-hectare area, a production capacity of 10.28 million t per year, and an overall productivity of 1368 kg/ha during 2020-21 as per the fourth advanced estimate (GoI, 2021). As pearl millet has wider adaptability particularly in the dry tracts, it offers ample opportunity to harvest food-grains even in harsh climatic conditions. In this way, it provides food security to the smallholders belonging to the fragile ecological regions. Further, intercropping can add the diversity and nutritional security to the food-basket to the millions of the farmers surviving with subsistence farming. Thus, pearl millet is playing a magnificent role in food and nutrition to the poorer section of the people and also contributing to the hunger-free world. The advantages of pearl millet in intercropping systems are narrated under the following subheadings.

9.3.1 Yield Advantages in Intercropping with Pearl Millet

For measuring the yield advantages in any intercropping system, LER is considered as the first choice where productivity of crop mixture is evaluated in respect to their pure stands. As stated earlier, the LER value with more than 1.0 indicated yield advantages in the intercropping system. The following table (Table 9.6) presented the LER values of some pearl millet based intercropping systems where yield benefits were clearly reflected.

Table 9.6. LER values of pearl millet-based intercropping systems

Intercropping systems	Ratio	LER	References
Pearl millet + Groundnut	1:3	1.54	Ghosh, 2004
Pearl millet + Pigeon pea	2:1	1.50	Ansari and Rana, 2012
Pearl millet + Sesbania	2:1	1.38	Dhaka *et al.*, 2014
Soybean + pearl millet	2:1	1.50	Layek *et al.*, 2014
Pearl millet + green gram	2:1	1.29	Renu *et al.*, 2018

The ATER is another concept where efficiency of an intercropping system is evaluated temporally and the time duration is considered for the period the crops in combination occupied the field.

Table 9.7. ATER of pearl millet-based intercropping systems

Intercropping systems	Ratio	ATER	References
Pearl millet + Sesbania	1:1	1.09	Dhaka *et al.*, 2014
Pearl millet + soybean	Binary mixture	1.48	Iqbal *et al.*, 2018
Pearl millet + green gram	2:1	1.25	Renu *et al.*, 2018

Intercropping combination seems to be advantageous when the base crop equivalent yield is higher in mixed stand than in the sole crop. Table 9.8 illustrates the benefits of intercropping in terms of pearl millet equivalent yield (PEY).

Table 9.8. Pearl millet equivalent yield (PEY) of intercropping system

Intercropping systems	Ratio	PEY (t ha^{-1})	References
Pearl millet + Pigeon pea	2:1	4.82	Ansari and Rana, 2012
Pearl millet + cluster bean	1:1	5.24	Bana *et al.*, 2016
Pearl millet + green gram	2:1	4.25	Renu *et al.*, 2018
Pigeonpea + Pearl millet	1:2	2.001	Vajjaramatti and Kalaghatagi, 2018
Pearl millet + moth bean	1:1	3.13	Kiroriwal *et al.*, 2012

9.3.2 Other Benefits

In many cases, pearl millet recorded superior growth attributing characters such as plant height, number of tillers etc. in mixed stand compared to their pure stands and it was obtained due to greater level of complementarity among the crop species. Similarly, higher grain yield of pearl millet was recorded by the researchers in the intercropping system (Ansari *et al.,* 2012). Legumes in intercropping system with pearl millet are known to check soil erosion, nutrient loss from the soil and dep2ression of weeds (Aasha *et al.*, 2017). Also, the intercropping system with pearl millet are recognized as a more income generating cropping system (Girase *et al.*, 2007; Rani *et al.*, 2017; Rawat *et al.*, 2018).

9.4 Upland Cotton in Intercropping

India ranks first in the global cotton production map, but the productivity is comparatively less than the world average (Blaise and Kranthi, 2019). It is an industrial crop and supports around 60 million people in the value chain in India. In our country, four species of cotton are grown (namely, *Gossypium herbaceum*, *G. arboreum*, *G. hirsutum* and *G. barbadense*). Among them, *Gossypium hirsutum* occupies 97% of cotton area in India and the species is also commonly known as American cotton and upland cotton. In India, upland cotton is grown as a rainfed crop in vast areas. In our country, about two-third of cotton is grown as a rainfed crop where intercropping can play a vital role in agricultural sustainability in cotton cultivation (Panda *et al.*, 2020). In general, a wide spacing is maintained in upland cotton that provides the scope to include intercrops spatially and temporally (Gitari *et al.* 2020). Further, inclusion of legumes in cotton-based intercropping systems results in an added advantage (Maitra and Ray, 2019).

9.4.1 Yield Advantages in Cotton-based Intercropping System

For quantifying the yield advantage of intercropping systems, some competitive functions are considered such as LER, ATER, base-crop equivalent yield and so on. In the following table, yield advantage in terms of LER in cotton-based intercropping systems has been shown (Table 9.9).

Table 9.9. LER values of upland cotton in intercropping systems

Intercropping systems	Ratio	LER	References
Upland cotton + Cowpea	1:1	1.46	Kumar *et al.*, 2017
Upland cotton + Cowpea	1:2	1.44	Rajpoot *et al.*, 2018
Upland cotton + Lady's finger	1:2	1.44	Rajpoot *et al.*, 2018
Upland cotton + red gram	6:1	1.04	Lateef Pasha *et al.*, 2020
Upland cotton + Sesame	1:1	1.08	Sathishkumar *et al.*, 2020

Like LER, the ATER also measures the efficiency of an intercropping system where time factor is considered indicating occupancy by individual crops in the mixed stand (Table 9.10).

Table 9.10. ATER of upland cotton-based intercropping systems

Intercropping systems	Ratio	ATER	References
Cotton + cowpea	1:1	1.13	Aasim *et al.* 2008
Cotton + cowpea	2:2	1.55	Mwamlima *et al.* 2016
Cotton + black gram	1:2	1.12	Vasavi and Sreerekha, 2017
Cotton + okra	1:2	1.20	Rajpoot *et al.* 2018

Intercropping combination is seems to be advantageous when the base crop equivalent yield is higher in intercropping than in the sole crop. Table 9.11 illustrates the advantage of intercropping in terms of cotton equivalent yield (PEY).

Table 9.11. Cotton equivalent yield (CEY) of intercropping systems

Intercropping systems	Ratio	CEY (kg ha^{-1})	References
Cotton + onion	1:1	2396*	Jayakumar and Surendran, 2016
Cotton + moth bean	1:2	714**	Patel *et al.*, 2017
Cotton + soybean	4:10	9174*	Turkhede *et al.*, 2017

* Seed cotton yield ** Lint yield

9.4.2 Multiple Benefits of Cotton Based Intercropping Systems

Intercropping is beneficial for its multiple benefits as recorded in different crop combinations (Maitra *et al.*, 2021). Yield enhancement, increase in productivity due to complementarity and economics of cotton cultivation was noted earlier (Maitra *et al.*, 2001). Sankaran and Balasubramanian (1982) recorded advantages of the cotton + legume intercropping system in weed population dynamics and suppression of weeds. Intercropping cotton with other crops was recorded as advantageous in terms of pest management (Suresh and Dason, 1996; Mote *et al.*, 2001).

9.5 Small Millets and Intercropping

Small millets are ancient food grains cultivated widely by small farmers in Africa and Asia because of their diversity, wider adaptability, nutritional value and suitability to combat adverse climatic conditions. In many regions, these are grown in resource-poor conditions where other commercial crops cultivation is not suitable. Among different small millets, finger millet or *ragi* (*Eleusine coracana* L. Gaertn) is widely grown. Besides, foxtail of Italian millet (*Setaria italica* L.), barnyard millet (*Echinochloa frumentacea* L.), little millet (*Panicum sumatrens* L.), *kodo* millet (*Paspalum scrobiculatum* L.), little millet (*Panicum sumatrense* L.), proso millet (*Panicum miliaceum* L.) and brown-top millet (*Brachiaria ramosa* L. Stapf; *Panicum ramosum* L.) are some small millets which are grown in small pockets in the world with local demand or regional significance. India contributes the major production of small millets (Rao *et al.*, 2013). The country grows small millets in about 0.7 million ha and average yield is 633 kg/ha (Anbukkani *et al.* 2017).

During the present time, small millets have regained importance in diet because of their nutritional quality and are considered as nutri-cereals (Banerjee and

Maitra, 2020). They are rich in minerals and fibres. The byproduct or stover is palatable to livestock and grain and fresh forage are also nutritious. India had the pride and heritage of small millets production, but during the post-Green Revolution era, focus was shifted to production of fine cereals, namely, rice, wheat and maize; and small millets were neglected. Besides, institutional support was also provided in favour of fine cereals. The most important cropping system of India is rice-wheat, and yield plateauing was noticed with the same. Moreover, the cropping system is considered as water-guzzler that created some associated problems such as soil salinity, water table depletion, heavy metal toxicity and other environmental issues. On the other hand, small millets are ecologically sound, hardy in nature, drought and heat tolerant and can use more CO_2; therefore, in the present consequence of climate change the crops have acquired their lost pride and importance (Passi and Jain, 2014).

Intercropping small millets with other crops (particularly legumes) has numerous benefits (Maitra, 2020; Gitari *et al*., 2020). As cultivation of small millets is adopted by smallholders and farmers adopt subsistence farming, food and nutritional security is important to them. Further, small millets are grown in resource-poor areas, soil fertility improvement through intercropping is an added advantage. As the crops remained comparatively neglected, researchers did not focus much on studying intercropping with small millets. However, there are research evidences on intercropping with finger millet, little millet and foxtail millet. Based on earlier studies, intercropping small millets under diverse conditions has been described as under.

9.5.1 Finger Millet in Intercropping

Intercropping with cereal and legume is a very common combination and it provides numerous advantages in terms of total productivity of crops, efficient use of available resources, soil fertility improvement and control of run-off of water and erosion and enhancement of diversity. The studies revealed that finger millet-legume intercropping was beneficial in terms of enhancement of productivity and profitability, efficient resource utilization, reduction in soil erosion and nutrient loss and improvement of soil fertility. The on-farm experiment carried out in Koraput of south Odisha indicated that intercropping finger millet with pigeon pea and black gram in replacement series reduced the yield of finger millet than pure stand (Dass and Sudhishir, 2010). However, the combined yield of finger millet and legumes was more and both the combinations of finger millet and red gram and black gram produced more finger millet equivalent yield (FMEY) than sole finger millet. The resource use efficiency was greater in intercropping as the highest LER value of 1.34 was recorded with intercropping of finger millet red gram (in ratios of 5:2 and 6:2). Intercropping finger millet with black gram reduced runoff (by 10.2%) and soil loss. Further, loss of nutrients such as N, P and K with soil erosion was also checked by growing the intercrops in contour due to coverage of ground area in contour by intercropping black gram in finger millet. Jakhar *et al.* (2015) worked on strip cropping of finger millet + groundnut and observed the above combination at 6:4 row ratio showed the highest value in terms of FMEY, net income and benefit: cost ratio. Intercropping of finger millet with legumes is a very common practice in India and finger millet + legumes combination registered higher productivity, comparatively less loss of yield due to biotic factors with greater resource and input use efficiency (Meena *et al.*, 2017). Moreover, as per the choice of legumes and proportions of crops in India, it was noted that finger millet + pigeon pea (8-10: 2) or finger millet + field bean (8: 1) is quite popular among the farmers of Karnataka and Tamil Nadu, but in Bihar, intercropping combination of finger millet and field bean (6: 2) was noted to perform well. For Garhwal region of Uttarakhand, finger millet + soybean (90:10 percent crop mixtures) and for Kolhapur region of Maharashtra finger millet + moth bean /black gram with row ratio of 4:1 was observed to be beneficial. Maitra *et al.* (2000) worked in the red and lateritic belt of West Bengal and opined that intercropping of finger millet in combination with pigeon pea and groundnut (4:1) resulted in more yield and resource use efficiency (as LER) when compared to pure stand of finger millet.

In Ethiopia, the mixed cropping of lupine and finger millet at the 50:100 and 75:100 seeding proportion had a greater yield advantage compared to sole cropping of either lupine or finger millet (Bitew, 2014). Manjunath and Salakinkop (2017) noted higher LER when soybean intercropped with finger millet at 2:1 and 4:2 proportions and the LER values were 1.45 and 1.47 which were more than unity, showing advantages of intercropping. Yadav (2018) highlighted greater resource use efficiency in terms of enhancement of the LER by finger millet and green gram or pigeon pea intercropping system and increase in productivity as finger millet equivalent yield that harnessed manifold. Finger millet when intercropped with haricot bean showed superiority to lupine (Bitew *et al.*, 2019) and the study revealed that intercropping of these legumes with finger millet could be synonymous to food and nutritional security of smallholders.

9.2.1 Foxtail Millet in Intercropping

Sufficient research was not conducted on intercropping in foxtail millet. However, Himasree *et al.* (2017) recorded that growing foxtail millet and red gram with a proportion of 5:1 resulted in higher LER, ATER, foxtail millet grain equivalent yield and benefit-cost ratio. Manjunath and Salakinkop (2017) noted more LER and benefit-cost ratio of intercropping foxtail millet and soybean (1:2 or 2:4). In another experiment, Manjunath *et al.* (2018) recorded that intercropping red gram + foxtail millet (1:2) registered more benefit cost ratio than pure stand of foxtail millet. Intercropping groundnut and foxtail millet (6:1) more LER and benefit-cost ratio than sole cropping of groundnut or foxtail millet (Shwethanjali *et al.*, 2018).

9.2.2 Little Millet in Intercropping

The practice of intercropping is common in little millet production. In a field experiment during *kharif* season conducted at Dharwad, Karnataka on alfisols clearly indicated that the little millet + pigeonpea intercropping system of 5:1 row proportion recorded more dry weight, length of ear and grain weight (Patil *et al.,* 2010). But the highest little millet equivalent yield (LMEY) was noted with 4:2 row proportion. Relay intercropping of little millet + horsegram recorded more LMEY than sole little millet alone. The intercropping combination of soybean + little millet (4:2) resulted in higher resource use efficiency as LER (1.39) compared to pure stands of soybean and B:C ratio was 1.95 which clearly indicated the advantage of the intercropping with little millet (Manjunath and Salakinkop, 2017). Sharmili and Parasuraman (2018) conducted an experiment in Tiruvannamalai district of Tamil Nadu during *kharif* season to study the impact of intercropping little millet with red gram and lablab bean on growth and productivity of crops. The study revealed that higher LMEY was noted in little millet + red gram with 6:1 row proportion followed by horse gram sequence. Intercropping combination of groundnut and little millet (6:1 row proportion) assured better resource use efficiency as LER (1.13) than pure stands of either groundnut or little millet and more B:C ratio of 2.16 (Shwethanjali *et al.*, 2018).

9.2.3 Other Millets in Intercropping

Research finding is meagre on intercropping of other millets, namely, proso millet, barnyard millet and brown top millet. However, on the basis of available literature, some information is presented here. Manjunath *et al.* (2018) mentioned that intercropping redgram + proso millet (1:2) resulted in more net return and B:C ratio over sole cropping. The grain yield of proso millet when intercropped with mung bean was improved by 6.8–37.3% compared with the pure stand of proso millet in China (Gong *et al.*, 2019) and this was achieved by delay in the proso millet flag leaf senescence in intercropping which helped in increasing grain yield of millet. Further, Milenkovic *et al.* (2019) reported that 1:1 ratio of soybean and proso millet in intercrop resulted in high biomass yield in Belgrade, Serbia. Further, Chapke *et al.* (2018) commented that proso millet + green gram (2:1) as a common combination of intercropping in India (Chapke *et al.*, 2018). Singh and Arya (1999) reported that the grain yield of barnyard millet was higher in pure stand compared to crop mixture, but combination of barnyard millet and soybean seed mixture of 90:10 percent recorded higher LER (1.45) and barnyard millet equivalent yield. Further, Chapke *et al.* (2018) mentioned that in Uttaranchal barnyard millet and rice bean (4:1) intercropping system is recommended and in Madhya Pradesh, intercropping of kodo millet and different legumes, namely, pigeon pea, green gram, black gram and soybean at a row ratio of 2:1 is commonly practiced.

In the chapter, benefits of intercropping with some major crops have been narrated. Intercropping system becomes advantageous when the crop components show a high level of complementarity. Moreover, proportion of individual species and duration also exert in expression of advantages. The above factors if not properly considered, intercropping may not show the desired benefits. There is a future need to study more crop combinations with different proportions suitable for different agroclimatic zones for development of ideal cropping systems targeting agricultural sustainability.

References

Aasha, M. B., Eltayeb, A. H., Abusin, R. M. A., Khalil, N. A. 2017. *Effects of intercropping pearl millet with some legumes on Striga hermonthica emergence. SSRG Int. J. Agric. Environ. Sci.* **4**(6): 64-72.

Aasim, M., Umer, E.M. and Karim, A. 2008. Yield and competition indices of intercropping cotton (*Gossypium hirsutum* L.) using different planting patterns. *Ankara üniversitesi ziraat fakültesi*, **14**(4): 326–333.

Ae, N., Arihara, J., Okada, K., Yoshihara, T. and Johansen, C. 1990. Phosphorus uptake by pigeon pea and its role in cropping systems of the Indian subcontinent. *Sci.* **248**:477-480.

Ahmad, A., Riaz, A., Mahmood, N. and Muhammad, M. 2006. Competitive performance of associated forage crops grown in different forage sorghum-legume intercropping systems. *Pak. J. Agric. Sci.* **43**(1-2):25-31.

Ahmad, M., R. Ahmad. and A. Rehman. 2007. Crowding stress tolerance in maize hybrids. Economic and Business Review. The Daily Dawn, Lahore, Pakistan. October 17- 22, 2007. p. 3.

Alfa, J. A. (2015). Performance of sorghum/soyabean intercrop as influenced by cultivar and row arrangement in the Northern Guinea Savanna Vegetational Belt of Nigeria. *Int. J. Sci. Basic App. Res.* **24**(5):184-203.

Amanullah, Khalid, S., Khalil, F. and Imranuddin. 2020. Influence of irrigation regimes on competition indexes of winter and summer intercropping system under semi-arid regions of Pakistan. *Sci Rep.***10**:8129 | https://doi.org/10.1038/s41598-020-65195-7.

Anbukkani, P., Balaji, S. J. and Nithyashree, M. L. 2017. Production and consumption of minor millets in India- A structural break analysis. *Ann. Agric. Res.* **38**:1-8.

Ansari, M. A. and Rana, K. S. 2012. Effect of transpiration suppressants and nutrients on productivity and moisture use efficiency of pearl millet (*Pennisetum glaucum*)-pigeonpea (*Cajanus cajan*) intercropping system under rainfed conditions. *Ind. J. Agric. Sci.* **82**(8):676-680.

Arshad, M. and Ranamukhaarachchi, S.L. 2012. Effects of legume type, planting pattern and time of establishment on growth and yield of sweet sorghum-legume intercropping. *AJCS*, 6(**8**):1265-1274.

Baker, C.; Modi, A.T.; Nciizah, A.D. Sweet sorghum (*Sorghum bicolor*) performance in a legume intercropping system under weed interference. *Agron.* **11**:877. https://doi.org/10.3390/agronomy11050877.

Bana, R. S., Pooniya, V., Choudhary, A. K., Rana, K. S. and Tyagi, V. K. 2016. Influence of organic nutrient sources and moisture management on productivity, biofortification and soil health in pearl millet (*Pennisetum glaucum*) + clusterbean (*Cyamopsis tetragonaloba*) intercropping system of semi-arid India. *Ind. J. Agric. Sci.* **86**(11): 1-418.

Banerjee, P. and Maitra, S. 2020. The role of small millets as functional food to combat malnutrition in developing countries. *Ind. J. Nat. Sci.* **10**(60): 20412-20417.

Behera, M. K. 2017. Assessment of the state of millets farming in India. *MOJ. Eco. Environ. Sci.* 2:16 20.

Bitew, Y. 2014. Influence of small cereal intercropping and additive series of seed proportion on the yield and yield component of lupine (*Lupinus* Spp.) in north western Ethiopia. *Agric Forest Fisheries.* **3**(2):133-14. doi: 10.11648/j.aff.20140302.23.

Bitew, Y., Alemayehu, G., Adego, E. and Assefa, A. 2019. Boosting land use efficiency, profitability and productivity of finger millet by intercropping with grain legumes. *Cogent Food Agric.* **5**:1702826. doi.org/10.1080/23311932.2019.1702826.

Blaise, D. and Kranthi, K. R. 2019. Cotton production in India. In: Cotton production, Eds. Jabran, K. and Chauhan, B.S. pp.193–215. DOI: 10.1002/9781119385523.ch10.

Chaithanya, J., Arvadiya, L.K. and Yallanagouda, M. 2021. Yield potential, land equivalent ratio and economic viability of summer sorghum (*Sorghum bicolor* L.) under sole crop and intercropping systems in south Gujarat condition. 2021. *Pharma Innov. J.* **10**(5): 753-755.

Chapke, R.R., Prabhakar, Shyamprasad, G., Das, I.K. and Tonapi, V.A. 2018. Improved millets production technologies and their impact. Technology Bulletin, ICAR-Indian Institute of Millets Research, Hyderabad 500 030, India:84p. ISBN: 81-89335-69-3.

Chen, G., Kong, X., Gan, Y., Zhang, R., Feng, F., Yu, A., Zhao, C., Wan, S. and Chai, Q. 2018. Enhancing the systems productivity and water use efficiency through coordinated soil water sharing and compensation in strip intercropping. *Sci. Rep.***8**:10494.

Choudhary, V. K. 2014. Suitability of maize-legume intercrops with optimum row ratio in mid hills of eastern Himalaya, India. *SAARC J. Agri.*, **12**(2): 52-62.

Crème, A., Rumpel, C., Gastal, F., De la Luz Mora Gil, M. and Chabbi, A. 2016. Effects of grasses and a legume grown in monoculture or mixture on soil organic matter and phosphorus forms. *Plant Soil.* **402**(1):117-128.

Crews, T. and Peoples, M. 2004. Legume versus fertilizer sources of nitrogen: ecological tradeoffs and human needs. *Agric. Ecosyst. Environ.***102**(3):279–297.

Dass, A. and Sudhishir, S. 2010. Intercropping in fingermillet (*Eleusine coracana*) with pulses for enhanced productivity, resource conservation and soil fertility in uplands of Southern Orissa. *Ind. J. Agron.*, **55** (2):89-94.

Dereje, G, Adisu, T., Mengesha, M. and Bogale, T. 2016. The influence of intercropping sorghum with legumes for management and control of striga in sorghum at Assosa Zone, Benshangul Gumuz Region, Western Ethiopia, East Africa. *Adv. Crop Sci. Tech.* **4**: 238. doi: 10.4172/2329-8863.1000238.

Dhaka, A. K., Pannu, R. K., Kumar, S., Poddar, R., Singh, B. and Dhindwal, A. S. (2014). Performance of seed crop of prickly *sesban* or *dhaincha* (*Sesbania aculeata*) when intercropped with pearl millet (*Pennisetum glaucum*). *Ind. J. Agron.* **59**(1): 70-75.

Duvvada, S.K. and Maitra, S. (2020). Sorghum-based intercropping system for agricultural sustainability. *Ind J. Nat. Sci.* **10**(60): 20306-20313.

FAOSTAT (2018). Production of sorghum. Available at: http://www.fao.org/faostat /en/#data/QC/visualize (Accessed on May 16, 2020).

FAOSTAT, 2022. Data, Crops and Livestock Products, Food and Agriculture Organization of United Nations, https://www.fao.org/faostat/en/#data/QCL (Accessed 23 January, 2022).

Gebremichael, A., Bekele, B. and Tadesse, B. 2019. Evaluation of the effect of sorghum-legume intercropping and its residual effect on yield of sorghum in Yekiworeda, Sheka zone, Ethiopia. *Int. J. Agril. Res. Innov. Tech.* **9**(2): 62-66. DOI: 10.3329/ijarit.v9i2.45412.

Ghanbari, A., Dahmardeh, M., Siahsar, B.A. and Ramroudi, M. 2010. Effect of maize (*Zea mays* L.) - cowpea (*Vigna unguiculata* L.) intercropping on light distribution, soil temperature and soil moisture in and environment. *J. Food Agr. Environ.*, **8**:102-108.

Ghosh, P. K. 2004. Growth, yield, competition and economics of groundnut/cereal fodder intercropping systems in the semi-arid tropics of India. *Field Crops Res.* **88**(2-3), 227-237.

Girase, P.P., Sonawane, P.D. and Wadile, S.C. 2007. Effect of pearl millet (*Pennisetum glaucum*) based intercropping system on yield and economics of pearl millet on shallow soils under rainfed conditions. *Int. J. Agric. Sci.* **3**(2): 192-193.

Gitari, H. I., Nyawade, S. O., Kamau, S., Karanja, N. N., Gachene, C. K. K., Raza, M. A., Maitra, S., Schulte-Geldermann, E. 2020. Revisiting intercropping indices with respect to potato-legume intercropping systems. *Field Crops Res.* **258**:107957.

GOI. (2021). Agricultural Statistics at a Glance, Ministry of Agriculture & Farmers Welfare Department of Agriculture, Cooperation & Farmers Welfare Directorate of Economics and Statistics, New Delhi. pp. 56, https://foodprocessingindia.gov.in/uploads/publication/Agricultural-statistics-at-a-Glance-2020.pdf (Accessed on August 29, 2021).

Gong, X.W., Liu, C.J., Ferdinand, U., Dang, K., Zhao, G. Yang, P. and Feng, B.L. 2019. Effect of intercropping on leaf senescence related to physiological metabolism in proso millet (*Panicum miliaceum* L.). *Photosynthetica.* **57** (4):993-1006. DOI: 10.32615/ps.2019.112.

Hiebsch, C.K. 1978. Interpretation of yields obtained in crop mixture. Abstracts of American Society of Agronomy, Madison, Wisconsin, pp.41.

Himasree, B., Chandrika, V., Sarala, N.V. and Prasanthi, A. 2017. Evaluation of remunerative foxtail millet (*Setaria italica* L.) based intercropping systems under late sown conditions. *Bull. Env. Pharmacol. Life Sci.* **6**(3):306-308.

ICAR-IIMR Annual report, 2020. ICAR-Indian Institute of Maize Research, PAU Campus,Ludhiana-141004. https://iimr.icar.gov.in/publications-category/annual-reports/ICAR-IIMR Director's report, 2019-20.

Iqbal, M. A., Hamid, A., Hussain, I., Siddiqui, M. H., Ahmad, T., Khaliq, A. and Ahmad, Z. 2019. Competitive indices in cereal and legume mixtures in a South Asian environment. *Agron. J.* **111**(1), 242-249.

Jakhar, P., Adhikary, P. P., Naik, B. S. and Madhu, M. 2015. Finger millet – groundnut strip cropping for enhanced productivity and resource conservation in upland of Eastern Ghats of Odisha. *Ind. J. Agron.* **60**(3):365-371.

Jan, R., Saxena, A., Jan, R., Khanday, M. and Jan, R. 2016. Intercropping indices and yield attributes of maize and black cowpea under various planting pattern, *Bioscan*, **11**(2):00-00 (Supplement on Agronomy).

Jayakumar, M. and Surendran, U. 2016. Intercropping and balanced nutrient management for sustainable cotton production. *J. Plant Nutri.* **40**(5): 632–644.

Jiao, N.Y., Zhao, C., Ning, T. Y., Hou, L.T., Fu, G.Z., Li, Z. J. and Chen, M. C. 2008. Effects of maize-peanut intercropping on economic yield and light response of photosynthesis. *Chinese J. Applied Ecol.* **19**: 981-985.

Joshi, B. K., Bhatta, M. R., Ghimire, K. H., Khanal, M., Gurung, S. B., Dhakal, R. and Sthapit, B. R. (2017). Released and promising crop varieties of mountain agriculture in Nepal (1959-2016). Pokhara, Nepal: LI-BIRD/Bioversity International, 207 p.

Kahsay, R., Mekuriaw, Y. and Asmare, B. 2021. Effects of inter-cropping lablab (*Lablab purpureus*) with selected sorghum (*Sorghum bicolor*) varieties on plant morphology, sorghum grain yield, forage yield and quality in Kalu District, South Wollo, Ethiopia. *Tropical Grasslands-Forrajes Tropicales*, **9**(2):216–224.

Keating, B.A. and Carberry, P.S. 1993. Resource capture and use in intercropping: Solar radiation. *Field Crops Res.* **34**:273-301.

Kermah, M., Franke, A.C., Samuel, A.N., Benjamin D.K.A., Abaidoo, R.C and Giller, K.E. 2017. Maize- grain legume intercropping for enhanced resource use efficiency and crop productivity in the Guinea savanna of northern Ghana, *Field Crops Res*. **213**:38-55.

Khan, M.A.H., Sultana, N., Akter, N., Zaman, M. S. and Islam, M. R. 2018. Intercropping garden pea (*Pisium sativum*) with maize (*Zea mays*) at farmers field Bangladesh. *J. Agril. Res.* **43**(4): 691-702.

Khan, Z. R., Midega, C. A. O., Hassanali, A., Pickett, J. A. and Wadhams, L. G. 2007. Assessment of different legumes for the control of *Striga hermonthica* in maize and sorghum. *Crop Sci.* **47**:730-734.

Kimou, S. H., Coulibaly, L.F., Koffi, B.Y., Toure, Y., Dede, K. J. and Kone, M. 2017. Effect of row spatial arrangements on agromorphological responses of maize (*Zea mays* L.) and cowpea (*Vigna unguiculata* L.). *Afr J. Agric. Res.* **12**(34): 2633-2641.

Kiroriwal, A., Yadav, R. S. and Kumawat, A. 2012. Weed management in pearlmillet based intercropping system. *Ind. J. Weed Sci.* **44**(3): 200–203.

Kumar, R., Turkhede, A.B., Nagar, R.K. and Nath, A. 2017. Effect of different intercrops on growth and yield attributes of American cotton under dryland condition. *Int. J. Curr. Microbiol. Appl. Sci.* **6**(4): 754-761.

Latheef Pasha. M. D., Sridevi, S., Ramana, M.V., Reddy, R.R., Goverdhan, P.M. and Jagan Mohan Rao. P. 2020. Yield, Economics and Cropping Indices of Cotton + Pigeon Pea Inter Cropping under Varying Row Ratios in Rained Conditions. *Int. J. Curr. Microbiol. App. Sci.* **9**(02): 1464-1472. doi: https://doi.org/10.20546/ijcmas. 2020.902.169.

Layek, J., Shivakumar, B. G., Rana, D. S., Munda, S., Lakshman, K., Das, A. and Ramkrushna, G. I. 2014. Soybean–cereal intercropping systems as influenced by nitrogen nutrition. *Agron J.* **106**(6), 1933-1946.

Litsinger, J. A. and Moody, K. 1976. Integrated pest management in multiple cropping system. In: Ppaendick, R. I., Sanchez, P. A. and Triplett, G. B. (eds.), Special Publication Number: 27, American Society of Agronomy, Madison, Wisconsin, USA, pp. 293-316.

Machado, S. 2009. Does intercropping have a role in modern agriculture? *J. Soil Water Conserv.* **64**(2):55A-57A.

Maitra, S. 2020. Intercropping of small millets for agricultural sustainability in drylands: A review. *Crop Res*. **55** (3 & 4):162-171.

Maitra, S. and Ray, D. P. (2019). Enrichment of biodiversity, influence in microbial population dynamics of soil and nutrient utilization in cereal-legume intercropping systems: a review. *Int. J. Biores. Sci.* **6**:11-19.

Maitra, S. and Ray, D. P. 2019. Enrichment of biodiversity, influence in microbial population dynamics of soil and nutrient utilization in cereal-legume intercropping systems: A Review. *Int. J. Biores. Sci.* **6**:11–19, doi:10.30954/2347- 9655.01.2019.3.

Maitra, S., Ghosh, D.C., Sounda, G., Jana, P.K., and Roy, D.K. 2000. Productivity, competition and economics of intercropping legumes in finger millet (*Eleusine coracana*) at different fertility levels. *Ind. J. Agric. Sci*. **70**:824–828.

Maitra, S., Samui, R.C., Roy, D.K. and Mondal, A.K. 2001b Effect of cotton based intercropping system under rainfed conditions in Sundarban region of West Bengal. *Indian Agric.* **45**(3-4): 157–162.

Maitra, S., Shankar, T. and Banerjee, P. (2020). Potential and advantages of maize-legume intercropping system, In: Maize - production and use; Hossain, A., Ed.; Intechopen, London, United Kingdom. doi:10.5772/intechopen.91722.

Maitra, S., Hossain, A., Brestic, M., Skalicky, M., Ondrisik, P., Gitari, H., Brahmachari, K., Shankar, T., Bhadra, P., Palai, J.B., et al. Intercropping-A low input agricultural strategy for food and environmental security. *Agron.* 2021, **11**: 343. https://doi.org/10.3390/agronomy11020343.

Mallikarjun, Koppalkar, B. G., Desai, B. K., Basavanneppa, M. A., Narayana Rao, K. and Swamy, M. 2018. Performance of pigeonpea (*Cajanus cajan*) intercropping as influenced by row ratios and nutria-cereal crops. *Int. J. Curr. Microbiol. App. Sci.* **7**(06): 2653- 2658. doi: https://doi.org/10.20546/ijcmas.2018.706.314.

Manasa, P., Maitra S., Reddy M. D. 2018. Effect of summer maize-legume intercropping system on growth, productivity and competitive ability of crops, *Int. J. Manage. Tech. Engg.* **8** (12), 2871-2875.

Manasa, P., Maitra, S. and Barman, S. 2019. Yield attributes, yield, competitive ability and economics of summer maize-legume intercropping system. *Int. J. Agric. Env. Biotechnol.* **13**(1): 33-38, DOI: 10.30954/0974-1712.1.2020.16.

Manjunath, M. and Vajjaramatti and Kalaghatagi, S.B. 2018. Performance of pigeonpea and millets in intercropping systems under rainfed conditions. *J. Farm Sci.* **31**(2):199-201.

Manjunath, M.G. and Salakinkop, S.R. 2017. Growth and yield of soybean and millets in intercropping systems. *J. Farm Sci.* **30**(3):349-353.

Mead, R., and Willey, R.W. 1980. The concept of a "land equivalent ratio" and advantages in yields from intercropping. *Exp Agric.* **16**:217-228.

Meena, D. S., Gautam, C., Patidar, O. P., Singh, R., Meena, H. M., Vishwajith, G. Prakash, V. G. 2017. Management of finger millet based cropping systems for sustainable production. *Int. J. Curr. Microbiol. App. Sci*. **6**(3):676-686. DOI:10.20546/ijcmas. 2017.603.078.

Milenkoviæ, M., Simiæ, M., Brankov, M., Milojkoviæ–Opsenica, D., Kresoviæ, B. and Dragièeviæ, V. 2019. Intercropping of soybean and proso millet for biomass production. *J. Process Energy Agric*. **23**(1): 38-40.

Mote, U.N., Patil, M.B. and Tambe, A.B. 2001. Role of intercropping in population dynamic of major pest of cotton ecosystem. *Ann. Plant Protec. Sci.* **9**: 36–40.

Mwamlima, L.H., Kabambe, V.H., Mhango, W.G. and Nyirenda, G.K.C. 2016. Effects of intercropping on growth and yield of intercropped cotton (Gossypium hirsutum) and cowpea (Vigna unguiculata L. Walp) in Malawi. *Agric. Sci. Res. J.* **6**(12): 303–312.

Narayanaswamy, P., Ganghadharan, K., Chandrasekharan, G., Velazhagan, R. and Karunanidhi, K. 1988. Proc. of National Workshop on Pests and Diseases, Tamilnadu Agricultural University, 16-18 September, 1988.

Nickel, J. L. (1973). Pest situation in changing agricultural system: a review. *Bull. Entomol. Soc. Amer*. **54**:76-86.

Panda, S. K., Maitra, S., Panda, P., Shankar, T., Pal, A., Sairam, M. and Praharaj, S. 2021. Productivity and competitive ability of rabi maize and legumes intercropping system. *Crop Res*. **56**(3-4):98-104; DOI:10.31830/2454-1761.2021.016.

Panda, S.K., Pritam Panda, P., Pramanick, B., Shankar, T., Subhashisa Praharaj, S., Saren, B.K., Harun I Gitari, H.I., Koushik Brahmachari, K., Akbar Hossain, A. and Maitra,S. (2020) Advantages of cotton based intercropping system: A review. *Int. J. Biores. Sci*. **7**(2): 51-57.

Passi, S. J. and Jain, A. 2014. *Millets: The nutrient rich counterparts of wheat and rice*. Government of India: Press Information Bureau, (Accessed: 29 August, 2021).

Patel, D. G., Patel, C. K., Singh, R. N. and Patel, N. I. 2017. Intercropping study in Bt. cotton under rainfed condition of Kutch region of Gujarat *(Gossypium hirsutum* L.). *Int. J. Sci. Environ. Tech.* **6**(6): 3484–3488.

Patil, N. B., Halikatti, S. I., Sujay, Y. H., Prasanna Kumar, B. H., Topagi, S. C. and Pushpa, V. 2010. Influence of intercropping on the growth and yield of little millet and pigeonpea. *Int. J. Agric. Sci.* **6**(2):573-77.

Rajpoot, S.K., Rana, D.S. and Choudhary, A.K. 2018. *Bt*-cotton–vegetable-based intercropping systems as influenced by crop establishment method and planting geometry of *Bt*-cotton in Indo-Gangetic plains region, *Curr. Sci.* **115**(3): 516–522.

Rani, S., Goyat, R. and Soni, J. K. 2017. Pearl millet (*Pennisetum glaucum* L.) intercropping with pulses step towards increasing farmer's income under rainfed farming: A review. *Pharma Innov. J.* **6**(10): 385-390.

Rao P., Parthasarathy and Basavaraj G. 2013. Status and prospects of millet utilization in India and global scenario. *In:* Millets: Promotion for Food, Feed, Fodder, nutritional and environment security, proceedings of global consultation on millets promotion for health & nutritional security. Society for Millets Research, ICAR Indian institute of millets research: Hyderabad, pp. 197–209.

Rawat, B. Kamboj, N. K., Kumar, R. and Gupta, G. 2018. Effect of nutrient management practices on yield and economics of pearl millet based intercropping systems under rainfed situations. *Ann. Agric. Res.* New Series **39** (2): 1-6.

Raza, M. A., Khalid, M. H. B., Zhang, X., Feng, L. Y., Khan, I., Hassan, M. J., Ahmed, M., Ansar, M., Chen, Y. K., Fan, Y. F., Yang, F. and Yang, W. 2019. Effect of planting patterns on yield, nutrient accumulation and distribution in maize and soybean under relay intercropping systems. *Sci Rep.* **9**:4947.

Renu, Kumar, A. and Kumar, P. 2018. Performance of advance pearl millet hybrids and mungbean under sole cropping and intercropping systems under semi-arid environment. *J. Pharmacog. Phytochem.* **7**(2): 1671-1675.

Reza, Z. O. (2012) Evaluation of Quantitative and Qualitative Traits of Forage Sorghum and Lima Bean under Different Nitrogen Fertilizer Regimes in Northern Guinea Savannah, Nigeria. *Nigerian J Basic Appl. Sci.* **19**(2):253-259.

Sankaran, S. and Balasubramanian, N. 1982. Inter cropping of cotton. *Cotton Dev.* **12**: 23–23.

Satheeshkumar, N., Thukkaiyannan, P., Ponnuswamy, K. and Santhi, P. (2011). Effect of sowing and weed management methods and intercrops on weed control and grain yield of sorghum under intercropping situation. *Crop Res.***41**:46-51.

Sathishkumar, A., Srinivasan, G., Subramanian, E. and Rajesh, P. 2020. Intercrops and weed management effect on productivity and competition indices of cotton. *Ind. J. Weed Sci.* **52**(2): 153–159; DOI: 10.5958/0974-8164.2020.00028.3.

Sharmili, K. and Parasuraman, P. 2018. Effect of little millet-based pulses intercropping in rainfed conditions. *Int. J. Chem. Studies.* **6**(6):1073-1075.

Shwethanjali, K. V., Kumar Naik, A. H., Basavaraj Naik, T. and Dinesh Kumar, M. 2018. Effect of groundnut + millets intercropping system on yield and economic advantage in Central Dry Zone of Karnataka under rainfed condition. *Int. J. Curr. Microbiol. App. Sci.* **7**(9):2921-2926. doi: 10.20546/ijcmas.2018.709.363.

Singh, D., Mathimaran, N., Boller, T. and Kahmen, A. 2020. Deep-rooted pigeon pea promotes the water relations and survival of shallow-rooted finger millet during drought – Despite strong competitive interactions at ambient water availability. *PLoS ONE.***15**: 0228993.

Singh, R.V. and Arya, M.P.S. 1999. Effect of seed ratio on barnyard millet (Echinochloa frumentacea) based mixed cropping system, *Ind. J. Agron.* **44**(1):51-55.

Somu, G., Meena, N., Shashikumar, C., Navi, S., Druvakumar, M., Kanavi, M. S. P. and Krishna Kishore, R. 2020. Evaluation of sorghum based intercropping system for yield maximization in sorghum, *Ind. J. Pure App. Biosci.* **8**(1), 145-149. doi: http://dx.doi.org/10.18782/2582-2845.7982.

Suresh, S. and Dason, A.A. 1996. Effect of intercropping and time of sowing on cotton leaf hopper and bollworm. *Madras Agric. J.* **83**: 56–57.

Tanwar, S.P.S., Rao, S.S., Regar, P.L., Shiv, D., Praveen, K., Jodha, B.S., Santra, P., Rajesh, K. and Rameshwar, R. 2014. Improving water and land use efficiency of fallow-wheat system in shallow Lithic Calciorthid soils of arid region: Introduction of bed planting and rainy season sorghum–legume intercropping. *Soil Till Res.***138**:44–55.

Teferi, T.J., Gashaw, M., Jemberu, T., Adebabay, A., Tadesse, T., Medhin, Z. 2020. Effect of sorghum/pulses intercropping on the productivity of farmlands in the moisture deficit areas of Belesa district, north west Ethiopia. *Int. J. Adv. Life Sci. Technol.* **4**(1):19-32. DOI: 10.18488/journal.72.2020.41.19.32.

Tripathi, A. K. and Kushwaha, H. S. 2014. Production potential of sorghum (*Sorghum bicolor*) intercropped with pigeon pea (*Cajanus cajan*) under varying fertility levels in the rainfed environment of Bundelkhand region. *Ann. Agric. Res. New Series*. **35**(4):373-378.

Tripathy S. K. 2019. Quality Protein Maize (QPM): A Way Forward for Food and Nutritional Security. *Genom. Appl. Biol.* **10**(2): 10-19, doi: 10.5376/gab.2019.10.0002.

Turkhede, A. B., Nagdeve, M. B., Karunakar, A. P., Gabhane, V. V., Mohod, V. D. and Mali, R. S. 2017. Diversification in cotton-based cropping system under mechanization in rainfed condition of Vidarbha of Maharashtra, India. *Int. J. Curr. Microbiol. Applied Sci.* **6**(9): 2189–2206.

USDA (2019). Agricultural Research Service. Food data. Available at: https://fdc.nal.usda.gov/fdc-app.html#/?query=sorghum (Accessed on August 29, 2020).

Vajjaramatti, M. M. and Kalaghatagi, S. B. 2018. Performance of pigeon pea and millets in intercropping systems under rainfed conditions. *J Farm Sci.* **31**(2), 199-201.

Vasavi, M. and Sreerekha, M. 2017. Effect of legume intercropping on competition indices and returns of Bt cotton. *J. Cotton Res. Dev.* **31**(2): 279–282.

Vesterager, J.M., N.E. Nielsen and H. Hogh-Jensen, 2008. Effects of cropping history and phosphorus source on yield and nitrogen fixation in sole and intercropped cowpea-maize systems. *Nutr. Cycl. Agroecosyst.* **80**: 61-73.

Wahbi, S., Prin,Y., Thioulouse, J., Sanguin, H., Baudoin, E., Maghraoui, T., Oufdou, K., Le Roux, C., Galiana, A., Hafidi, M., Duponnois, R. (2016). Impact of wheat/*faba* bean mixed cropping or rotation systems on soil microbial functionalities. *Front. Plant Sci.***7**:1364.

Willey, R. W. 1979. Intercropping— Its importance and research needs. Part-2. Agronomy and research approaches. *Field Crops Abst.* **32**:73-81.

Willey, R. W., and Reddy, M. S. 1981. A field technique for separating above-and below- ground interactions in intercropping: an experiment with pearl millet/groundnut. *Exp. Agric.* **17**:257-264.

Yadav, R.C. 2018. Racy Nature (Land and Water)-SIMM (System of Intensification of Minor Millets) combo endowed with eco-zero weeding towards quantum mechanics. *Acta Sci. Agric.* **2**(7):61-73.

Yogesh, S., Halikatti, S. I., Hiremath, S. M., Potdar, M. P., Harlapur, S.I. and Venkatesh, H. 2014. Light use efficiency, productivity and profitability of maize and soybean intercropping as influenced by planting geometry and row proportion, *Karnataka J. Agric. Sci.* **27**(1): 1-4.

Colour Plates

Chapter 1: Overview of Intercropping

Figure 1.1. Additive series of intercropping groundnut in paired row maize in south Odisha

Chapter 4: Advantages of Intercropping

Figure 4.1. (A) Intercropping sesame and groundnut **(B)** Maize and cluster bean

Chapter 7: Alley Cropping

Figure 7.1. Alley cropping of perennial tree and annual crop

Full grown canopy of *Leucaena* alley during lean period

Prunned *Leucaena* alley two weeks after maize sowing

Alley cropping of maize at full grown stage in *Leucaena*

Figure 7.2. Schematic drawing showing management of *Leucaena* and maize alley cropping

Chapter 8: Legumes in Intercropping and Soil Improvement

Figure 8.1. Cowpea (3 rows) intercropped in paired row maize (2 rows)

Figure 8.2. Fenugreek intercropped in sweet corn

Chapter 9: Performance of Crops in Intercropping

Figure 9.1: Maize and chickpea intercropping system; **(A)** paired row maize + 3 rows chickpea (2:3) and **(B)** uniform row maize + single row chickpea (1:1)